DISASTER EDUCATION

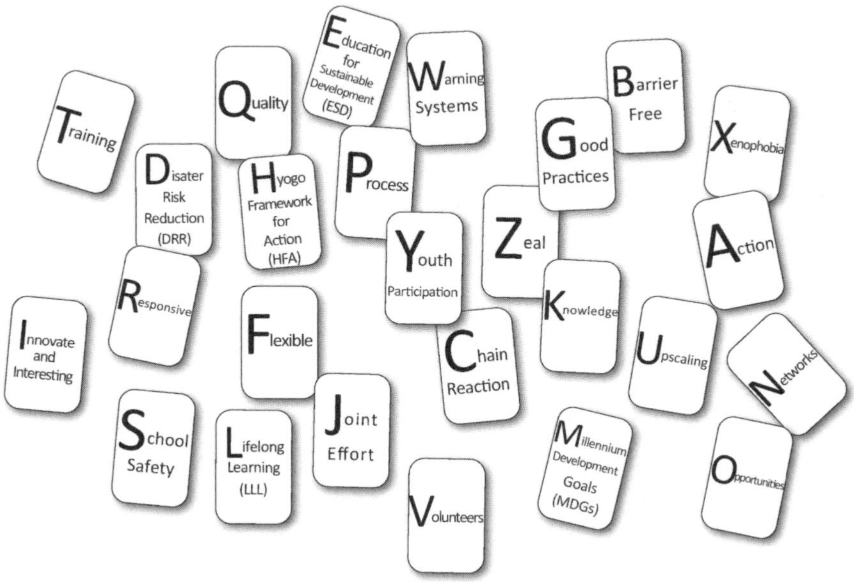

COMMUNITY, ENVIRONMENT AND DISASTER RISK MANAGEMENT VOLUME 7

DISASTER EDUCATION

EDITED BY

RAJIB SHAW

Graduate School of Global Environmental Studies, Kyoto University, Kyoto, Japan

KOICHI SHIWAKU

Earthquake Disaster Mitigation Research Center, Kobe, Japan

YUKIKO TAKEUCHI

Graduate School of Global Environmental Studies, Kyoto University, Kyoto, Japan

Emerald

United Kingdom – North America – Japan
India – Malaysia – China

Emerald Group Publishing Limited
Howard House, Wagon Lane, Bingley BD16 1WA, UK

First edition 2011

British Library Cataloguing in Publication Data
A catalogue record for this book is available from the British Library

ISBN: 978-0-85724-737-7
ISSN: 2040-7262 (Series)

CONTENTS

LIST OF CONTRIBUTORS

Glenn Fernandez	International Environment and Disaster Management Laboratory, Graduate School of Global Environmental Studies, Kyoto University, Kyoto, Japan
Qi Ru Gwee	International Environment and Disaster Management Laboratory, Graduate School of Global Environmental Studies, Kyoto University, Kyoto, Japan
Fuad Mallick	Postgraduate Programs in Disaster Management, BRAC University, Dhaka, Bangladesh
Farah Mulyasari	International Environment and Disaster Management Laboratory, Graduate School of Global Environmental Studies, Kyoto University, Kyoto, Japan
Rajib Shaw	International Environment and Disaster Management Laboratory, Graduate School of Global Environmental Studies, Kyoto University, Kyoto, Japan
Koichi Shiwaku	Earthquake Disaster Mitigation Research Center (EDM), National Research Institute for Earth Science and Disaster Prevention (NIED), Kobe, Japan
Yukiko Takeuchi	International Environment and Disaster Management Laboratory, Graduate School of Global Environmental Studies, Kyoto University, Kyoto, Japan

LIST OF EDITORS

Rajib Shaw
Graduate School of Global Environmental Studies, Kyoto University, Kyoto, Japan

Koichi Shiwaku
Earthquake Disaster Mitigation Research Center (EDM) National Research, Institute for Earth Science and Disaster Prevention (NIED), Kobe, Japan

Yukiko Takeuchi
Graduate School of Global Environmental Studies, Kyoto University, Kyoto, Japan

EDITORS' PROFILE

Rajib Shaw is an Associate Professor in the Graduate School of Global Environmental Studies of Kyoto University, Japan. He worked closely with the local communities, NGOs, governments and international organization, including United Nations, especially in the Asian countries. He is currently the Chair of the United Nations Asia Regional Task Force for Urban Risk Reduction. His research interests are: community-based disaster risk management, climate change adaptation, urban risk management, and disaster and environmental education.

Koichi Shiwaku is a Research Fellow of Earthquake Disaster Mitigation Research Center (EDM), National Research Institute for Earth Science and Disaster Prevention (NIED), Japan. He worked closely with the local community, school teachers and students, NGOs, governments, universities, and other related organizations. His main working fields are mainly Asian countries including disaster-affected areas. His research interests are community-based disaster management, disaster education, and capacity building.

Yukiko Takeuchi is an Assistant Professor in the Graduate School of Global Environmental Studies of Kyoto University, Japan. She has extensive experiences in risk communication and gender studies in the field of disaster risk reduction. She has worked in several developing countries in the Asian region. Working with the school children and local communities, she has developed different innovative tools for disaster education and community-based risk communication. She has also written extensively in different journals and book chapters.

BRIEF INTRODUCTION TO THE SERIES

COMMUNITY, ENVIRONMENT AND DISASTER RISK MANAGEMENT

This series connects academic research to field practice, strengthening the links between the environment, disaster, and community. The series will be developed on field evidences and community practices, and thus will provide specific guides to professionals who are grounded in rigorous academic analysis. The series has a specific focus on community-based disaster risk management, urban environmental management, human security, water community, risk communication, climate change adaptation, climate disaster resilience, and community-based practices.

BRIEF INTRODUCTION TO THE VOLUME

Disaster education has its different dimensions, from school, family, to community education. Education is a process that needs to be embedded at different levels of management and practice to collectively reduce risk. While school education is the foundation of the knowledge cycle, for effective knowledge use a link between school and community education is required. Education is linked to enhanced awareness, and a key reflection of education is seen in terms of actions. Disaster education is practiced in different countries in different forms, based on the local socioeconomic and cultural contexts. This book is an attempt to describe and demonstrate different aspects of disaster education in an easy-to-understand form with practical field experiences.

PREFACE

Disaster education is a much-discussed topic in risk reduction literatures. Education itself is always welcomed, and there have been different frameworks and conventions on education, whether it is a right-based approach or a part of development perspective or an environmental issue. Disaster education is considered as a crosscutting issue, which needs to be incorporated in different existing educational frameworks. As obvious, disaster education deals with practical matters, and cannot be a stand-alone school or university curriculum. There needs to be a good balance between the curriculum, and extra-curriculum activities, and in-school, outside-school activities. Outside-school activities can be of different types, in family and in community. Therefore, the disaster education is considered to link the school, family, and education.

One of the most important aspects of disaster education is its process, approaches, and tools. The process and approaches need to be innovative, in most cases student led, and tools need to link the theory to practice. Some of the tools are lecture or presentation oriented, but some tools are more linked to observations or experiential learning. Educational governance or policy is an important issue which makes the key pillar of the sustainability in the educational sector. The higher education is an important and crucial issue, which often gets less priority, but builds the core of the professionals and future generations in the respective field. With these concepts and ideas, this book consists of eight chapters which addressed some of the above issues through illustrative examples. It is difficult to measure the impacts of education, since it is a long-term investment. A modest attempt is done to understand the impacts of some of the educational approaches and tools like neighborhood watching.

Through our own education and research experiences, we strongly felt the need of a consolidated compilation on disaster education, the current literature on which is rather scattered and diverse. This book is the modest attempt to do so. Disaster education is an evolving subject. New ideas, methods, and tools will be generated over time. We hope that this book will be a good trigger to the future research in the subject. We will be delighted if the readers consider the book useful.

Rajib Shaw
Koichi Shiwaku
Yukiko Takeuchi
(*Editors*)

CHAPTER 1

DISASTER EDUCATION: AN INTRODUCTION

Rajib Shaw, Yukiko Takeuchi, Qi Ru Gwee and Koichi Shiwaku

ROLE OF EDUCATION IN DISASTER RISK REDUCTION

It has been widely acknowledged that education takes on a pivotal role in reducing disasters and achieving human security in the attempt to achieve sustainable development. Previous experiences have shown positive effects of education in disaster risk management. Children who have been taught about the phenomenon of disasters and how to react to those situations have proved to be able to respond promptly and appropriately, thereby warning others and protecting themselves during times of emergencies. One of the classic examples illustrating the power of knowledge and education is the story of the 10-year-old British schoolgirl, Tilly Smith, who warned the tourists to flee to safety moments before the Indian Ocean tsunami engulfed the coast, saving over 100 tourists' lives in 2004. She had recognized the signs of an approaching tsunami after learning about the phenomenon in her geography lessons at school, just weeks before visiting Thailand (UN/ISDR, 2006a). Although the United Kingdom is not a tsunami-prone country and the schoolgirl did not have any previous experiences, with the knowledge acquired at school, she was able to save the lives of many.

Disaster Education
Community, Environment and Disaster Risk Management, Volume 7, 1–22

ISSN: 2040-7262/doi:10.1108/S2040-7262(2011)0000007007

Here is another example of how knowledge, in this case, indigenous knowledge transferred from generations to generations, helped save the lives of thousands of people. During the same Indian Ocean tsunami, only seven people were killed out of the total population of about 83,000 on the Simeulue Island, located off the coast of Sumatra that was only 100 km away from the epicenter of the massive earthquake that caused the catastrophic event. The community had previously experienced a tsunami in 1907, and, since that time, the knowledge of tsunamis had been transmitted from generation to generation through cultural practices such as songs and poems. The simple knowledge "if you feel the earth shake, then immediately get away from the seashore" was embedded in lore and thereby saved many lives. This is a lesson of the power and simplicity of an experience learned and preserved (UN/ISDR, 2006b). These are just one of the few examples showing how formal and informal education had helped raised the awareness among individuals, thereby saving their lives as well as others in the community.

In this context, this chapter provides a basic introduction on disaster education. The chapter is divided into three main parts: the first part provides a summary of formal, non-formal, and informal education. The second part provides a literature review on the existing concepts of education, and the final part provides the outline of the book.

FORMAL, NON-FORMAL, AND INFORMAL EDUCATION

When talking about education, it can be broadly classified into three modes, namely formal, non-formal, and informal. The International Standard Classification of Education (ISCED) (1997) and the European Commission (EC) give the following definition of formal, non-formal, and informal education:

Formal education (or initial education or regular school and university education): Education provided in the system of schools, colleges, universities, and other formal educational institutions that normally constitutes a continuous "ladder" of full-time education for children and young people, generally beginning at age five to seven and continuing up to 20 or 25 years of age. In some countries, the upper parts of this "ladder" are constituted by organized programs of joint part-time employment and part-time participation in the regular school and university system: such programs have come to

be known as the "dual system" or equivalent terms in these countries (UNESCO, 1997). Typically provided by education or training institutions, with structured learning objectives, learning time, and learning support, it is intentional on the part of the learner and leads to certification (EC, no date).
Non-formal education: Any organized and sustained educational activities that do not correspond exactly to the above definition of formal education. Non-formal education may therefore take place both within and outside educational institutions, and cater to persons of all ages. Depending on country contexts, it may cover educational programs to impart adult literacy, basic education for out-of-school children, life skills, work skills, and general culture. Non-formal education programs do not necessarily follow the "ladder" system, and may have differing duration (UNESCO, 1997). It is not provided by an education or training institution and typically does not lead to certification. However, it is intentional on the part of the learner and has structured objectives, times, and support (EC, no date).
Informal education: Results from daily activities related to work, family life, or leisure. It is not structured and usually does not lead to certification. In most cases, it is unintentional on the part of the learner (EC, no date).

Table 1 provides a summary of the definitions of the three modes of education.

Coombs and Ahmed (1974) defined the three modes of education as follows: formal education is "institutionalized, chronologically graded and hierarchically structured educational system, spanning lover primary school and the upper reaches of the university"; non-formal education is "any organized, systematic, educational activity carried on outside the framework of the formal system to provide selected types of learning to particular subgroups in the population, adults as well as children," and informal education is "the lifelong process by which every person acquires and accumulates knowledge, skills, attitudes and insights from daily experiences and exposure to the environment" (Norland, 2005).

The major difference between formal and non-formal education is the influence of the government on the sponsorship (La Belle, 1981), while the major difference between non-formal and informal education is the deliberate instructional and programmatic emphasis present in non-formal education but absent in informal education (Norland, 2005). In practice, formal, non-formal, and informal may exist simultaneously, sometimes in concert with one another. For example, the school provides formal education, and at the same time fosters non-formal education through extra-curricular activities. Throughout the different phases of an individual's lifetime, he or

Table 1. Summary of the Definitions of Formal, Non-Formal, and Informal Education.

	Formal	Non-formal	Informal
ISCED (1997)	• *Location* In the system of schools, colleges, universities, and other formal educational institutions • *Structure* Normally constitutes a continuous "ladder" of full-time education Generally full-time education Cater mainly to aged 5–25 years May be in the form of organized programs of joint part-time employment and part-time participation in the regular school and university system	• *Location* In any organized and sustained educational activities, both within and out of educational institutions • *Structure* Do not necessarily follow the "ladder" system May have differing duration Cater to all ages Educational programmes are country and local- context based	
European Commission	• *Location* Typically provided by education or training institutions • *Structure* Structured objectives, times and support, and intentional on the part of the learner Leads to certification	• *Location* Not provided by education or training institutions • *Structure* Structured objectives, times and support, and intentional on the part of the learner Typically does not lead to certification	• *Location* Daily activities, anywhere • *Structure* Not structured No certification Unintentional on the part of the learner
Coombs and Ahmed, (1974); Norland (2005)	• *Structure* Institutionalized, chronologically graded, and hierarchically structured educational system, spanning from primary school to the university	• *Structure* Any organized, systematic, educational activity out of the formal system, providing selected types of learning to particular subgroups in the population	• *Structure* Lifelong process, accumulates from daily experiences and exposure to the environment

It should be noted that although there are international definitions, there may be slight variations based on the country and cultural contexts.

she comes into contact with these different modes of education. For example, most young children prior to school age are engaged primarily in the informal mode. Subsequently, as they enter school, they start to receive formal education, and there may be involvement of non-formal education in school (e.g., extra-curricular activities organized by school) and out of school (e.g., television, community activities).

It should be noted that although there are international definitions, there may be slight variations based on the country and cultural contexts.

Smyth (1996) pointed out that education should be treated as a sustained learning experience for everyone throughout life. Childhood is perhaps the age that has the most important memories in a person's life and in which an individual's personality takes form. Most adults retain and unconsciously use information commencing from upbringing experiences in the earlier stages of their lives. Almost everything that children learn is provided in different ways by the outside world. It should be noted that children are very emotional and sensitive about everything they learn and can easily be attracted or repelled by any topic. Thus, education should not be presented to them with a sense of doom but rather by understanding the natural world (Talero, 2004). Furthermore, children learn better when they focus on local issues. Local environments should be the basis for curricula with six- through nine-year-olds, and only after they are able to think in an analytical manner can they learn in a global way (Sobel, 1995). In addition, education programs should be viewed in a continuous and progressive perspective. A study conducted during a four-week period in Costa Rica by Vaughan, Gack, Solorazano, and Ray (2003) revealed that if children are guided in a proper way, parents and other adults could also benefit from them, suggesting the transfer of knowledge from children to parents in a way from classroom to the community. From these examples, we can see the influence early education has on an individual, and thus education should be a continuous and progressive process. If well-guided, it could also be transferred from an individual to another.

Formal education certainly plays a major role; however, there are many other channels through which learning takes place such as at home, with the family, within the community and peer groups, at the workplace and recreation, public, and media. Modern communications have provided information for the growing public demand of related information. With proper conduct, media can offer good communication tools that can be used as educational aids to reduce the gap between scientific knowledge and civic awareness (Talero, 2004). Production of community maps has been proven to be a successful tool in developing a cost-effective yet efficient way to increase community awareness. In a publication by Denis Wood titled *Memory, Love, Distortion, Power: What Is a Map*, it was suggested that what is communicated with maps reflects one's relationship to the territory one inhabits; thus maps can be produced in a participatory way where members of the community (including children) contribute with the information and ideas, creating a "sense of inclusion"; this way, the map becomes an excellent tool both as an educational aid to enhance awareness and as a public participation tool that helps unite the community (Talero, 2004). A study was conducted to examine the effectiveness of direct mail video campaigns and was found to receive good acceptance by the public. The audio and visual character made it easier for video to deliver information and captivate audiences more readily as compared to printed materials (Talero, 2004). These are just some examples of non-formal education that have been proven to help in the learning process.

The relative influence of formal, non-formal, and informal education varies with time and circumstances, and different people respond to them differently. But they are all important. Formal, non-formal, and informal education should be partnered together, if possible, to achieve best results. Hofstein and Rosenfeld (1996) argued that integration of informal learning experiences within the formal school curriculum enriches the learning process. La Belle (1981) also suggested the contribution of non-formal education as a complement and supplement to formal education. Eshach (2007) explained that whereas schools have become increasingly anachronistic, museums have retained "the potential to engage students, to teach them, to stimulate their understanding, and most important, to help them assume responsibility for their own future learning." In Chapter 36 of Agenda 21, on "Promoting Education, Public Awareness and Training," it was stated that "while basic education provides the underpinning for any environment and development education, the latter need to be incorporated as an essential part of learning" (UNEP, 1992). "Both formal and non-formal education is indispensable to change people's attitudes so that they have the capacity to assess and address their sustainable development concerns," suggesting the synergetic effect resulting from the combination of the various types of education.

Non-formal education, such as museums visits, may have their complementary influence on education. Interactive exhibits have been designed to achieve effectiveness in terms of behavior and learning outcomes by engaging visitors in more interactions (Lucas, 1999). However, Wymer (1991) suggested that although interactive exhibits may have the potential to spark interest in students, they should be handled carefully, or they may simply be gimmicks to attract crowds, providing little education. John Dewey (1859–1952), one of the preeminent educational theorists, wrote *Experience and Education*, which addresses fundamental issues relating to practice of museum education, and believed that experience is the basis for education. However, Dewey also warns that an enjoyable experience can lead to no particular learning or, worse, mis-educative experience (Anshacher, 1998). Furthermore, if teachers do not have explicit goals for these non-formal activities, it would be difficult to connect the experience to the classroom curriculum (Eshach, 2007). Thus, prior to organizing extra-curricular activities (i.e., non-formal education), teachers should first be clear of the purpose and/or outcome and prepare students accordingly to help them gain better understanding of the activity.

Informal education applies to situations in life that come about spontaneously, which could be from within the family, in the neighborhood, at school or workplace, etc. Solomon (2003) argued, "no one would deny the influence of home and families on the education of our children, and there is little doubt that a large part of all early social learning happens there." Resnick (1987) stated, "popular wisdom holds that common sense outweighs school learning for getting along in the world-that there exists a practical intelligence, different from school intelligence, that matters more in real life." The significance of spontaneous informal education should not be overlooked.

Now that we have understood the three modes of education, their role and complementary effects, the next section discusses the incorporation of disaster risk reduction into the various modes of education.

DISASTER EDUCATION

"Disaster education," "disaster risk education," and "disaster prevention education" are different expressions that essentially mean disaster risk reduction education. Since the International Decade of Natural Disaster Reduction (IDNDR) in the 1990s, significant public education efforts emerged in many nations, and various educational materials for school children and the general public have been produced. "Public awareness" and "education for disaster risk reduction" are gaining importance and have been widely advocated. Smyth (1996) suggested that campaigns and programs should be designed in a sustainable way to assist target groups with the acquisition of knowledge, skills, and attitudes that are necessary to solve actual and local issues. Public awareness in communities has been widely promoted as part of many national, regional, and international project and initiatives. Partnership efforts between different institutions, like government and nongovernmental organizations, private businesses, or educational entities, would be necessary.

The Hyogo Framework for Action (HFA) 2005–2015, priority for action 3, emphasizes the role of "Knowledge and Education" and highlights formal and non-formal education and awareness-raising as important components for disaster risk reduction. Following the adoption of HFA, various educational materials in the form of booklets, handbooks, textbooks, posters, activities, games, and practices were developed (UN/ISDR, 2006a). "Disaster Reduction, Education and Youth" was the UN World Disaster Reduction Campaign in 2000, and in 2006–2007, there was the UN/ISDR campaign on "Disaster Risk Reduction Begins at School," both campaigns addressing the emphasis of integrating disaster risk reduction into education. In the 2006–2007 "Disaster Risk Reduction Begins at School" campaign, UN/ISDR not only attempted to highlight the importance of integrating disaster risk reduction into formal education, but also emphasized the importance of community participation in order to achieve sustainability within the community (UN/ISDR, 2006a). Some of the lessons learnt from the experiences include the following: (i) education is a process for effective disaster reduction; (ii) knowledge, perception, comprehension, and actions are the four important steps; (iii) schools and formal education play an important role in knowledge development; (iv) family-, community-, and self-education are important for comprehension of knowledge and implementation of risk reduction actions; and (v) holistic education includes actions at local level, as well as its policy integration (Shiwaku, 2009). Furthermore, the campaign also promoted safe construction of school buildings (UN/ISDR, 2006a). In addition to providing education, school buildings could also serve as a temporary shelter for the community following disasters; thus, the safety of buildings is important to ensure the safety of students as well as continuation of education following disasters.

Petal (2008) suggested that the goal of developing "disaster-resilient communities" is widely understood to depend heavily on the success of disaster risk reduction education. She further explained that the integration of both formal and non-formal/informal education through school is one way of ensuring that these messages reach every family and community so that the learning can be sustained for future generations. Lidstone and Nielsen (1999) also suggested that while formal disaster education would remain a systematic and structured learning that progresses over time, the real-life context, reflective learning, and situated-learning characteristics of non-formal/informal education are equally important.

In terms of formal education, it is widely acknowledged that school plays an important role in raising awareness among students, teachers, and parents (Shaw & Kobayashi, 2001). The importance of disaster education at school is increasing because of the following reasons: (i) children are one of the most vulnerable sections of the society during a disaster; (ii) they represent the future; (iii) school serves as a community's central location for meetings and group activities; and (iv) effects of education can be transferred to parents and community (Shiwaku, 2009; UN/ISDR 2006a). With this in view, the school is regarded to play a crucial role in raising awareness among students, teachers, and parents as well as within the local community. Petal (2008) suggested that in terms of formal education for disaster risk reduction, formal curriculum integration may be introduced in the form of elective course or modules into the current existing courses and suggested several methods in achieve such integration.

While conventional education is largely dependent on classroom lectures, textbook lessons, and exercises, studies have shown that disaster education is more successful through experience-based and action-oriented learning. Petal (2009) argued that the mission of Disaster Risk Reduction (DRR) education, both for children and for adults, in all walks of life should serve to: (i) convey an understanding of the natural and environmental conditions and the human actions and inaction that lead to disaster, to stimulate changes in individual and group behavior, and (ii) motivate advocacy and raise expectation of social policy to reduce these threats. It was further emphasized that disaster education should not be merely teaching "natural hazards" or organizing "campaigns for risk awareness" but should be guiding people toward the discovery of their own solution and their own power. Indeed, early efforts in disaster education focused exclusively on hazards, teaching about volcanoes, earthquakes, and floods that happened in other places, to other people, and in other times (Lidstone & Nielsen, 1999). Lidstone (1996) mentioned that despite finding much to praise in the material available on disaster risk reduction education, behavioral changes among the students is unlikely, and suggested that instead of the current emphasis on the physical nature of potential disastrous events, schools should concentrate on the students' involvement in the context of disasters and promote a geographical education that encourages students to regard themselves living in a dynamic physical environment. Sharma (2008) pointed out that while teaching-based approaches are dependent on the content developed by academicians, learning approach is based on people finding out things from the environment or local sources; hence the learning approach catalyzes the learning process that remains dependent on the local environment and encourages cross-learning through sharing of stories, facts, and cultural approaches. The local sources and information, also known as indigenous knowledge, evolved over generations in the community are time tested in the local context, and thus are seen to be able to sustain for generations.

Ronan, Johnson, Daly, and Fairley (2001) carried out a survey of school children in New Zealand, and found that despite the fact that students who received disaster education may have raised risk awareness, their ability to differentiate and respond appropriately to disaster was weak. Another study on school disaster education in Nepal found that school disaster risk reduction education, which is based on lectures, can raise risk perception, but cannot enable students to understand the importance of pre-disaster measures and to take actual action for disaster reduction (Shiwaku, Shaw, Kandel, Shrestha, & Dixitn, 2007.) Shiwaku et al. (2007) pointed out that community plays an essential role in promoting students' actual actions for disaster risk reduction. Shaw, Shiwaku, Kobayashi, and Kobayashi (2004) carried out a comparative study of various modes of education on risk perception among school students in Japan. Results showed that while 70% of the students have been given education on disaster, and up to 80% have an understanding of risk-related issues, only 30%–40% actually practice any preparedness measures, suggesting that mere education does not translate into desired actions. Instead, community and family level of education has more direct influence on better disaster preparedness. In the same study, it was indicated that school disaster risk reduction education, coupled with self, family, and community education, can help a student develop a "culture of disaster preparedness" in their communities, and it was suggested that disaster education should involved non-formal activities that influence actions rather than mere knowledge. Paton and Johnston (2001) encouraged community and social psychological factors to be included in disaster education so as to facilitate the relationship between risk perception and risk reduction behavior. Paton (2005) mentioned that disaster education should be integrated with community development initiatives to increase resilience and facilitate self-help capacities so as to reduce reliance on external response and recovery resources. Pandey and Okazaki (2005) emphasized the importance of educating the community by ensuring dissemination of best practices in disaster risk management at the community level and formulation of integrated programs for sustainable development. By doing so, it can foster and sustain a capacity within the individual as well as the community to adapt to the vicissitudes that may occur.

Yoshida (2007) gave the example of a school setting and suggested that the activity "town-watching" be performed by the students and teachers in the school system. "Town-watching," a participatory technique used in community or neighborhood planning in order for residents to recognize problems as a group and put forward solution together, has been extended to dealing with disaster and safety related to physical issues such as safe or unsafe places and evacuation routes. Another example of a successful disaster education program is the Earthquake Safety Education program in Iran. The program includes integration of disaster education into the formal education and performs non-formal education activities, at the same time, involving students in the local context as well as engaging participation from the community. Iran, being located near the faults, has high risks of seismic hazards and has suffered several earthquakes resulting in severe economic losses in the past. As such, the government has initiated the "School Earthquake Safety Initiative," "School Safety Act," as well as the "Earthquake Safety Education in School" (International Institute of Earthquake Engineering and Seismology [IIEES], 2000; International Institute of Earthquake Engineering and Seismology [IIEES], 2006). The Earthquake Safety Education program is led by the Iranian Ministry of Education (MOE) in collaboration with the International Institute of Earthquake Engineering and Seismology (IIEES) as well as other public and civic organizations. Disaster lessons are integrated within science, geography, literature, and other curricula with preparedness and practice books designed for different grades and handbooks for teachers (Petal, 2008). Safety drills are one of the most commonly performed activities. Every year, the IIEES, Iranian Ministry of Education, National Committee for Natural Disaster Reduction, Iranian Red Crescent Society, Iran National Television and Radio, and other related national organizations organize these drills to prepare students and staff for appropriate and rapid responses during earthquakes. There is also a broadcast of an "Earthquake safety alarm" on national and local radio. Furthermore, a "School Earthquake Safety Council" involving parents and teachers voluntarily serves as a form of disaster risk reduction and preparedness effort at the individual and school levels (Petal, 2008).

In addition to education in preparing the students in the event of a disaster, education in emergencies and reconstruction is equally important. Tens of millions of children are deprived of education in countries around the world hit by conflict or natural disaster. In 2000, the World Education Forum, held in Dakar, pledged to "meet the needs of education systems affected by conflict, natural calamities and instability" and countries to work toward the objective of Education for All, and a session on "Education in situations of emergency and crisis" was conducted (Sinclair, 2002). Sinclair suggests that emergency education should use a community-based participatory approach, with emphasis on capacity building, teachers training, and developing curriculum that enriches life skills and supports long-term development of individual as well as the society (Tables 2 and 3).

Table 2. Summary of the Findings from Literature Review on Formal, Non-Formal, and Informal Education.

Citation	Approaches and/or Process	Tools	Findings from the activity
Vaughan et al. (2003), p. 12	• 4-week program, teacher-guided, 3rd-4th grade • Students requested to read 2–3 coloring books with their parents and color them. Both students and parents to sign upon completion • Participants to take pre- and post-test	• Coloring book • Lessons (2 h/week)	• Continuous, progressive influence of early education on individual was seen • Transfer of knowledge from children (classroom) to parents (community) was possible.
Talero (2004), p. 13	• Creating community map • Video delivered through the mail	• Map • Mail video	• Creating maps may act as public participation tool, enhance awareness and unite community. • Video via mail captivates audiences more readily
Hofstein et al. (1996), p. 13	• Research literature on various informal activities such as museum visits, field trips, community-based programs	• Literature review	• Integration of informal and formal could enhance learning interest
La Belle (1981), p. 13	• Research literature on various private and public, profit and nonprofit programs in the United States	• Literature review	• Non-formal education may be the used as the dominant educational mode in among the poor • Structured and systematic characteristics of non-formal education complements with formal education.

Lucas (1999), p. 14	• Study on class visit to science center • Pre-visit (3 weeks and a week before the visit) and post-visit interview conducted with the teacher-in-charge and the students. Observations during the visit recorded.	• Field visit	• Teacher-in-charge tried to establish effective bridge between the formal and non-formal education by displaying posters and exhibits made by students prior to visit • Students found exhibits interesting
Anshacher (1998), p. 14	• Literature review on *John Dewey's Experience and Education: Lessons for Museums*	• Literature review	• 'All genuine education comes about through experience but it does not mean that all experiences are equally educative' • 'An enjoyable experience can lead to no particular learning or, worse mis-educative experience'
Solomon (2003), p. 14	• Study on a British project – School-Home Investigations in Primary Science (SHIPS) Project • Questionnaires survey and interview conducted	• Literature review	• Parents were willing to participate in the activities • Children spoke more at home, reinforcing bonding at home
Ronan et al. (2001), p. 17	• Survey to assess children's level of awareness, risk perceptions, factual knowledge, physical preparedness, and psychological issues related to hazards. Children's prior exposure to (a) specific hazards and (b) educational programs designed to increase awareness, knowledge, and preparedness was assessed as well	• Hazard lessons in school	• Hazards-educated children had greater awareness of the most appropriate hazard-related protective behaviors • However, some behaviors were found to be incorrect or not endorsed in the school education. Educated children showed a general inability to differentiate when it comes to some secondary response behaviors

Table 2. *(Continued)*

Citation	Approaches and/or Process	Tools	Findings from the activity
Shiwaku et al. (2007), p. 17	• Applied Rohrman's awareness consisting of three levels: (i) risk-appraisal, (ii) decision for prevention action, and (iii) risk-reducing behavior	• School Earthquake Safety Program	• Community plays an essential role for promoting students' actual actions for disaster risk reduction
Shaw et al. (2004), p. 17	• Comparative study on various modes of education on risk perception – school, family, community, and self.	• Hazard lessons • Voluntary activities • Newspaper, TV	• Majority of students (805) acquired understanding to risks but only 30-40% put the knowledge into practice • Community and family education has more direct influence on disaster preparedness
Yoshida (2007), p. 18	• Town-watching activity organized by school with participation from parents and community members	• Town-watching	• Students gained more understanding of natural hazards and local environment
IIEES (2000 and 2006), pp. 19, 20	• Iran Government-Earthquake Safety Education Initiative, School Safety Act and Earthquake Safety Education in School program • Collaboration between Ministry of Education with IIEES as well as public and civic organizations • Disaster lessons integrated into curricula • Activities, local radio broadcast	• textbooks, teachers' guidebooks • Educational materials, posters • Safety drills	• Integrated, holistic approach which includes legislative basis (School Safety Act), government's commitment, multi-stakeholder involvement, formal and non-formal education integration, community participation.

Table 3. Summary of the Suggestions from Literature Review on Formal, Non-Formal and Informal Education.

Citation	Suggestions
Petal (2008), p. 17	• Integration of formal and non-formal/informal through school aids ensuring disaster education messages to reach every family and community. • Integration may be achieved by (i) curriculum integration, (ii) extracurricular integration, (iii) curriculum infusion, and (iv) stand-alone course
Lidstone (1996) and Lidstone et al. (1999), p. 15	• Instead of emphasis on the physical nature of disasters, disaster education should concentrate on the students' involvement in the context of disasters • While formal disaster education remains as systematic and structured learning, real-life context, reflective learning, and situated-learning characteristics of non-formal/informal education are equally important
Paton et al. (2001), p. 17	• Community and social psychological factors to be included in disaster education to facilitate the relationship between risk perception and risk reduction behavior
Paton (2005), p. 17	• Disaster education should be integrated with community development initiatives to increase resilience and enhance self-help capacities
Pandey et al. (2005), p. 17	• Educate the community through dissemination of best practices in disaster risk reduction and formulation of integrated programs so as to foster and sustain capacity within the individual and the community, thereby enhancing the resilience of the community

UNESCO defines "educational emergency" as a crisis situation created by conflicts or disasters that have destabilized, disorganized, or destroyed the education system and that require an integrated process of crisis and postcrisis support, recognize the importance of ensuring education continuity following disasters, and take the lead in promoting education as part of emergency response and for long-term recovery (UNESCO, 2008). Education can save and sustain lives, offer physical and psychosocial protection, and at the same time, give people hope for the future and serve as a channel that promotes safety and well-being. It is much cheaper to prevent. Had a fraction of the hundreds of millions of dollars raised to rebuild Haiti been invested in advance – for earthquake-safe infrastructure, and in educating communities in disaster preparedness and response – it would have prepared the people and saved them from the event (UNESCO, 2010a). The UNESCO International Institute for Education Planning (IIEP) addresses the three core aspects to education in emergencies, which include the following (i) preparing the planning for emergency – an education sector diagnosis is necessary to examine the risk and vulnerabilities of the education system to both human and manmade disasters, which in turn help in developing budgeting for contingency planning and capacity strengthening within the education system; (ii) responding to an emergency – including rapid assessment of education needs, coordination of education actors, and capacity gap analysis for reconstruction; and (iii) continuing education during and after an emergency – depending on the context, especially strategies in ensuring continued provision of education, e.g., temporary learning spaces as well as advocacy to donors, agencies, and private sectors on funding education (UNESCO, 2010a). In an effort to promote the education in emergencies, the *Guidebook for Planning Education in Emergencies and Reconstruction* was developed by IIEP, primarily targeting the staff of ministries of education, national, provincial, and district-level planners and managers, in disaster-affected areas. The guidebook consists of five sections, namely: (i) general overview, (ii) access and inclusion, (iii) teachers and learners, (iv) curriculum and learning, and (v) management capacity. It provides a comprehensive coverage of areas related to education in emergencies and reconstruction (UNESCO, 2010b). For example, in section 2 (access and inclusion) of the guidebook, it has been suggested that non-formal education, e.g., radio programs and temporary classes, could be used for providing emergency education due to the accessibility to resources. However, it also reminds the user of the guidebook the importance of education to be focused on the needs and concerns of the learners at all times.

In addition to education in preparing the students in the event of a disaster, education in emergencies and reconstruction is equally important. Tens of millions of children are deprived of education in countries around the world hit by conflict or natural disaster. In 2000, the World Education Forum, held in Dakar, pledged to "meet the needs of education systems affected by conflict, natural calamities and instability" and countries to work toward the objective of Education for All, and a session on "Education in situations of emergency and crisis" was conducted (Sinclair, 2002). Sinclair suggests that emergency education should use a community-based participatory approach, with emphasis on capacity building, teachers' training, and developing curriculum that enriches life skills and supports long-term development of individual as well as the society.

The HFA highlights the importance of "Knowledge and Education," and during the 12th UN/ISDR Inter-Agency Task Force on Disaster Reduction Meeting (IATF/DR), it was agreed that a cluster approach would best combine IATF/DR and partners' efforts and achievements on the issue, which consists of different UN agencies and international NGOs. The cluster aims to strengthen the network, create new partnerships, identify gaps, and focus on areas so as to achieve HFA goals though knowledge and education (UN/ISDR, 2005). Apart from UN agencies, a self-organizing, voluntary initiative known as the Coalition for Global School Safety and Disaster Prevention Education (COGSS), initiated since 2008, supports the widespread commitment for the implementation of the HFA priority for action 3, and engages in workgroups and ad hoc committees to advance shared objectives for safe schools and disaster education (COGSS, 2008). The coalition consists of individuals from organizations, coalitions, committees, task forces, national and local governments, public, civic and private sector organizations, agencies, departments or divisions, as well as practitioners, and provides information-sharing, coordination, guidance, expertise, leadership, advocacy, program development, policy analysis, and research to promote disaster education and safe schools (COGSS, 2008). The four main focus areas of the coalition are as follows: (i) disaster-resistant school infrastructure, (ii) disaster education in schools through formal curricula, (iii) disaster prevention education: extra-curricular and community-based, and (iv) school-based disaster management (COGSS, 2008).

Chapter 36 of Agenda 21 stated that "Education is critical for promoting sustainable development and in providing the capacity of the people to address environmental and development issues," which was highlighted at the World Summit on Sustainable Development (WSSD, 2002). Undoubtedly, as discussed in this section, education plays a significant role in enhancing disaster risk reduction knowledge and awareness and at the same time promotes sustainable development. Education can be formal, non-formal, and/or informal, and examples of each mode have been discussed. Although school education may provide the individual disaster knowledge, it alone is insufficient to either raise preparedness or motivate the individual to take actions on disaster risk reduction activities. Education is vital, as is the sharing of experience within and among the communities (UN/ISDR, 2007). In addition to classroom lecture, supplementation of non-formal, experience-based, and action-oriented learning activities, with the incorporation of indigenous knowledge and community participation, is seen to enhance the awareness and preparedness within individuals as well as the community, and is widely advocated in disaster risk reduction education.

Furthermore, consideration and planning of disaster education should not be limited to education on "normal" days but should also include education during emergency times. Last but not least, regular monitoring and evaluation is essential.

ABOUT THIS BOOK

This book has eight chapters. The first chapter by Shaw Takeuchi Gwee and Shiwaku is the introduction, which provides an overview of the practices and ideas of disaster education. The second chapter by Gwee, Shaw, and Takeuchi focuses on the educational policy and its perspective. In very few countries, there is a specific disaster education policy, although the national and international frameworks provide opportunities and ways to develop respective policies at the national and local government levels. Analyzing the available policies, the chapter provides an insight into the E-HFA (Education in Hyogo Framework for Action) as a governance and policy framework for disaster education. The third chapter by Shiwaku and Fernandez provides direction of school disaster education. Citing the example of Maiko High School of Japan, school curriculum analysis of Nepal, and youth participation in the Philippines, the chapter shows the importance of a balanced approach of curriculum and extra-curriculum education, and the activities in the school as well as outside the school. The fourth chapter by Takeuchi Mulyasari and Shaw provides examples of family and community education. The chapter categorizes the family and community based on rural and urban lifestyles, and points out that each type of family or community has its unique way of disaster education. The chapter gives examples from Japan, India, and Indonesia and provides the generic ways of family and community education by summarizing the commonalities and differences among the examples. The fifth chapter by Shaw, Mallick, and Takeuchi analyzes the high education prospects for disaster risk reduction. Drawing lessons from the sustainable development education, the chapter provides specific examples of networks of disaster education and climate change education. The chapter emphasizes the importance of filed-based research and interdisciplinary education process. The sixth chapter by Shiwaku and Fernandez provides innovative approaches of disaster education. Involvement with the local communities, school, and education department in Japan, participatory involvement of the local NGOs, communities, schools, and local governments in Nepal and school-based disaster education in Indonesia are some of the key approaches of disaster education. Neighborhood watching is considered as an important approach to involve the students with the local communities. The seventh chapter by Mulyasari, Takeuchi, and Shaw provides different examples of tools for disaster education. Some of the tools are by lectures, some through experiential learning, and some through presentations. The chapter lists different disaster education tools for the curriculum and non-curriculum activities. Finally, the eighth chapter provides an analysis of the future direction of disaster education. By introducing the Japanese word "Tsunagaru," meaning linking, the chapter emphasizes that the disaster education is all about linking different elements, from school to community, different disciplines, different stakeholders, human and environment, and time. Thus, the book provides different dimensions and ideas of disaster education and provides future directions.

ACKNOWLEDGMENTS

The authors acknowledge the support and discussion among the authors and other people in the International Environment and Disaster Management Laboratory in Kyoto University, which is highly acknowledged.

REFERENCES

Anshacher, T. (1998). John Dewey K experience and education: Lessons for museums. *Curator: The Museum Journal*, *41*, 36–49.

Coalition for Global School Safety and Disaster Prevention Education (COGSS and DPE). (2008). Governance framework.

Coombs, P. H., & Ahmed, M. (1974). *Attacking rural poverty: How non-formal education can help*. Baltimore: John Hopkins University Press.

Eshach, H. (2007). Bridging in-school and out-of-school learning: Formal, non-formal, and informal education. *Journal of Science Education and Technology*, *16*(2), 171–190.

European Commission. (no date). *Education and training. Validation of non-formal and informal learning*. Available at http://ec.europa.eu/education/lifelong-learning-policy/doc52_en.htm (Accessed on November 19, 2010).

Hofstein, A., & Rosenfeld, S. (1996). Bridging the gap between formal and informal science learning. *Studies in Science Education*, *28*, 87–112.

International Institute of Earthquake Engineering and Seismology (IIEES). (2000). *Public education*. Available at http://www.iiees.ac.ir/iiees/English/Publicedu/eng_publicedu.html (Accessed on November 20, 2010).

International Institute of Earthquake Engineering and Seismology (IIEES). (2006). *Case study: IRAN. School disaster risk reduction: Think globally, act locally*. Available at http://

www.iiees.ac.ir/iiees/English/Publicedu/school_safety_iran_case_study_davos.pdf (Accessed on November 20, 2010).

ISCED (1997). International Standard Classification of Education, UNESCO, Paris. Available at http://www.unesco.org/education/information/nfsunesco/doc/isced_1997.htm. Retrieved on February 21, 2011.

La Belle, T. J. (1981). An Introduction to the nonformal education of children and youth. *Comparative Education Review*, *25*(3), 313–329.

Lidstone, J. (1996). Disaster education: Where we are and where we should be. In: Lidstone, J. (Ed.), *International perspectives on teaching about hazards and disasters* (p. 3). Philadelphia, USA: Channel View Publications.

Lidstone, J., & Nielsen, S. (1999). Public education and disaster management: Is there any guiding theory? *Australian Journal of Emergency Management*, *13*(3), 14–19.

Lucas, K. B. (1999). One teacher's agenda for a class visit to an interactive science center. *Science Education*, *84*, 524–544.

Norland, E. (2005). The nuances of being 'non': Evaluating nonformal education programs and settings. *New Directions for Evaluation*, *108*, 6–12.

Pandey, B., & Okazaki, K. (2005). Community based disaster management: Empowering communities to cope with disaster risks. *Regional Development Dialogue*, *26*(2), 52–57.

Paton, D. (2005). *Community resilience: Integrating hazard management and community engagement*. In: Proceedings of the international conference on engaging communities, Queensland Government/UNESCO, Brisbane. Available at http://www.engagingcommunities2005.org/abstracts/Paton-Douglas-final.pdf (Accessed on April 21, 2010).

Paton, D., & Johnston, D. (2001). Disaster and communities: Vulnerability, resilience and preparedness. *Disaster Prevention and Management*, *10*(4), 270–277.

Petal M. (2008). *Concept note: Formal and informal education for disaster risk reduction*. Available at http://www.riskred.org/activities/ddredislamabad.pdfwww.riskred.org/activities/ddredislamabad.pdf (Accessed on November 20, 2010).

Petal, M. (2009). Education in disaster risk reduction. In: R. Shaw & R. R. Krishnamurthy (Eds), *Disaster management: Global challenges and local solutions* (pp. 285–320). Hyderabad, India: University Press.

Resnick, L. B. (1987). Learning in school and out. *Educational Researcher*, *16*(9), 10–13.

Ronan K.R., Johnson D.M., Daly M., & Fairley R. (2001). School children's risk perceptions and preparedness: A hazards education survey. *Australasian Journal of Disaster and Trauma Studies* 1. Available at http://www.massey.ac.nz/~trauma/issues/2001-1/ronan.htm (Accessed on April 21, 2010)

Sharma, A. (2008). *Effective content and methodology of distance learning in community based disaster management*. Doctorate's Thesis, Kyoto University.

Shaw, R., & Kobayashi, M. (2001). The role of schools in creating earthquake-safer environment. In OECD Workshop, Thessaloniki.

Shaw, R., Shiwaku, K., Kobayashi, H., & Kobayashi, M. (2004). Linking experience, education, perception and earthquake preparedness. *Disaster Prevention and Management*, *13*(1), 39–49.

Shiwaku, K. (2009). Essentials of school disaster education: Example from Kobe, Japan. In: R. Shaw & R. R. Krishnamurthy (Eds), *Disaster management: Global challenges and local solutions* (pp. 321–337). Hyderabad, India: Universities Press.

Shiwaku, K., Shaw, R., Kandel, R. C., Shrestha, S., & Dixit, A. (2007). Future perspective of school disaster education in Nepal. *Disaster Prevention and Management, 16*(4), 576–587.

Sinclair, M. (2002). *Planning Education in and after emergencies.* International Institute for Education Planning, UNESCO. Available at www.unesco.org/iiep/PDF/Fund73.pdf (Accessed on November 19, 2010).

Smyth, J. C. (1996). A national strategy for environmental education: An approach to a sustainable future?. *The Environmentalist, 16*, 27–35.

Sobel, D. (1995). Beyond Ecophobia; reclaiming the heart in nature education. *Orion Society Nature Literacy Series, 14*(4). Available at http://www.yesmagazine.org/issues/education-for-life/803 (Accessed on November 19, 2010).

Solomon, J. (2003). Home-school learning of science: The culture of homes, and pupils' difficult border crossing. *Journal of Research in Science Teaching, 40*(2), 219–233.

Talero, G. (2004). *Literature review environmental education and public awareness.* Available at http://worldfish.org/PPA/PDFs/Semi-Annual%20II%20English/2nd%20s.a.%20eng_F2.pdf (Accessed on November 19, 2010).

UN/ISDR. (2005). *Thematic cluster/platform on knowledge and education.* Available at http://www.unisdr.org/eng/task%20force/working%20groups/knowledge-education/knowledge-education.htm (Accessed on November 10, 2009).

UN/ISDR. (2006a). Newsletter ISDR Inform-Latin American and the Caribbean. No.13. Available at: http://www.eird.org/eng/revista/no_13_2006/art7.htm (Accessed on November 17, 2010).

UN/ISDR. (2006b). *World disaster reduction campaign. Disaster risk reduction begins at school.* Available at http://www.unisdr.org/eng/public_aware/world_camp/2006-2007/pdf/WDRC-2006-2007-English-fullversion.pdf (Accessed on January 28, 2010).

UN/ISDR. (2007). *Building disaster resilient communities. Good practices and lessons learned.* Available at http://www.unisdr.org/eng/about.../education-good-practices.pdfwww.unisdr.org/eng/about.../education-good-practices.pdf (Accessed November 17, 2010).

UNEP. (1992). *Agenda 21/Chapter 36: Promoting education, public awareness and training.* Available at http://habitat.igc.org/agenda21/a21-36.htm (Accessed on January 12, 2010).

UNESCO. (2008). *Education in emergencies: Preparedness, response, recovery.* Available at http://www.unesco.org/new/en/unesco/themes/post-conflict-and-post-disaster-responses/education-in-emergencies/

UNESCO. (2010a). *International institute for education planning.* Available at http://www.iiep.unesco.org/en/focus-on-education-in-emergencies/education-in-emergencies-an-introduction.html

UNESCO. (2010b). *International institute for education planning, guidebook for planning education in emergencies and reconstruction.* Available at http://www.iiep.unesco.org/fileadmin/user_upload/Cap_Dev_Technical_Assistance/pdf/Guidebook/Guideboook.pdf (Accessed on December 10, 2010).

UNESCO, E. (1997). *International standard classification of education ISCED 1997* (Available at: http://www.unesco.org/education/information/nfsunesco/doc/isced_1997.htm. (Accessed on November17, 2010)). Paris: UNESCO.

Vaughan, C., Gack, J., Solorazano, H., & Ray, R. (2003). The effect of environmental education on schoolchildren, their parents, and community members: A study of intergenerational and intercommunity learning. *Journal of Environmental Education, 34*(3), 12.

World Summit on Sustainable Development (WSSD). (2002). Available at www.world summit2002.org/ (Accessed on December 10, 2010).

Wymer, P. (1991). Never mind the science, feel the experience. *New Scientist*, *132*(1789), 53.

Yoshida, Y. (2007). *Study on effective and sustainable community disaster education through town watching in Saijo City*. Master's Thesis. Kyoto University

CHAPTER 2

DISASTER EDUCATION POLICY: CURRENT AND FUTURE

Qi Ru Gwee, Rajib Shaw and Yukiko Takeuchi

INTRODUCTION

The importance of education in disaster risk reduction has been emphasized in several international agendas, frameworks, conferences, as well as UN programs. Chapter 36 of Agenda 21, on "Promoting Education, Public Awareness and Training" stated, "Education, including formal education, public awareness and training, should be recognized as a process by which human beings and societies can reach their fullest potential" (UNEP, 1992). Furthermore, the UN/ISDR System Thematic Cluster/Platform on Knowledge and Education argued that "Education for disaster risk reduction is an interactive process of mutual learning among people and institutions. It encompasses far more than formal education at schools and universities, and involves the recognition and use of traditional wisdom and local knowledge for protection from natural hazard" (UN/ISDR, 2005). In the 2006 Review of the Role of Education and Knowledge in Disaster Risk Reduction, Professor Ben Wisner commented, "Education, knowledge and awareness are critical to building the ability to reduce losses from natural hazards, as well as the capacity to respond to and recover effectively from extreme natural events when they do, inevitably, occur" (Wisner, 2006). The Second Asian Ministerial Conference on Disaster Risk Reduction (2007, India) urged governments to make school safety and the integration of

Disaster Education
Community, Environment and Disaster Risk Management, Volume 7, 23–44

ISSN: 2040-7262/doi:10.1108/S2040-7262(2011)0000007008

disaster risk reduction into school curricula a priority on the national agenda (UN/ISDR, 2007a). The Third Asian Ministerial Conference on Disaster Risk Reduction (2008, Malaysia) recognized education as an essential contribution to effective implementation of disaster risk reduction and concrete impact in terms of shifts in behaviors at the local level, where communities are most vulnerable to disasters (UN/ISDR, 2008). Last but not least, the UNESCO Education for Sustainable Development (ESD) program emphasized that "Education is the primary agent of transformation toward sustainable development, increasing people's capacities to transform their visions for society into reality" (UNESCO, 2005a).

The goal of developing "disaster-resilient communities" is widely understood to depend heavily on the success of disaster education, and the integration of both formal and non-formal/informal education through school is one way of ensuring that these messages reach into every family and community so that the learning can be sustained into the future generations (Petal, 2008). There still remain several challenges on educational policy, since there is not a single policy that covers different sectors of educational issues. Recognizing the significance of education in disaster risk reduction, suitable policy is necessary to integrate disaster risk reduction into education. This chapter provides an overview of the ongoing initiatives promoting the integration of disaster risk reduction in education, and introduces new concept of disaster policy.

INTERNATIONAL INITIATIVES

This section describes the evolution of processes at international levels that are related to disaster education. Broadly, this can be categorized as ESD (Education for Sustainable Development), EFA (Education for All), and Hyogo Framework for Action (HFA).

In 1989, the UN General Assembly designated the second Wednesday of October the International Day for Natural Disaster Reduction, which had been observed annually during the International Decade for Natural Disaster Reduction, 1990–1999. In 2001, the General Assembly decided to maintain the observance of the International Day for Natural Disaster Reduction as a vehicle to promote a global culture of natural disaster reduction, including disaster prevention, mitigation, and preparedness. In September 2000, world leaders came together to adopt the UN Millennium Declaration, committing their nations to a new global partnership to achieve eight anti-poverty goals by their 2015 target date, known as the

Millennium Development Goals (MDG), and to "achieve universal primary education" was among the eight target goals. In 2002, the Millennium Project was commissioned by the UN Secretary-General to develop a concrete action plan for the world to achieve the MDG. In the same year, the UN Millennium Campaign was initiated to inspire people in the world to take action in supporting the MDG (UN, 2010).

The ESD advocated by the UNESCO aims to help people develop the attitudes, skills, and knowledge required to make decisions for their own benefit and that of others, now and in the future, and to act upon these decisions. ESD supports five fundamental types of learning to provide quality education and foster sustainable human development – learning to know, learning to be, learning to live together, learning to do, and learning to transform oneself and society (UNESCO, 2005a). Learning to know aims to acquire knowledge, values, and skills for respecting and searching for knowledge and wisdom. Learning to be aims to acquire knowledge, values, and skills for personal and family well-being. Learning to live together aims to acquire knowledge, values, and skills for international, intercultural, and community cooperation and peace. Learning to do aims to acquire knowledge, values, and skills for active engagement in productive employment and recreation.

The four major aspects focused in ESD are as follows: (i) improving quality of basic education, (ii) reorienting educational programs, (iii) developing public understanding and awareness, and (iv) providing training (UNESCO, 2005b). In December 2002, the UN General Assembly put in place a UN Decade of Education for Sustainable Development (DESD), spanning from 2005 to 2014, which aims to promote the following in the educational scheme: (i) interdisciplinary and holistic learning rather than subject-based learning; (ii) values-based learning; (iii) critical thinking rather than memorizing; (iv) multimethod approaches: word, art, drama, debate, etc.; and (v) participatory decision-making and locally relevant, rather than national, information (UNESCO, 2005b). The primary aim is to encourage governments to consider the inclusion of measure to implement the DESD in their respective education systems and national development plans with four key objectives: (i) facilitating networking and collaboration among stakeholders in ESD, (ii) fostering greater quality of teaching and learning of environmental topics, (iii) supporting countries in achieving their MDG through ESD efforts, and (iv) providing countries with new opportunities and tools to reform education. UNESCO is designated to lead the Decade and seeks to integrate the principles, values, and practices of sustainable development into all aspects of education and learning, in order to address the social, economic, cultural, and environmental problems faced in the twenty-first century. Implementation focused on the following seven building blocks: (i) advocacy and vision-building, (ii) consultation and ownership, (iii) partnership and networks, (iv) capacity-building and training, (v) research and innovation, (vi) information and communication technologies, and (vii) monitoring and evaluation (UNESCO, 2005b).

The HFA 2005–2015 prepared by the UN/ISDR emphasizes the role of formal and non-formal education and awareness-raising as important components for disaster risk reduction. Following the adoption of HFA in 2005, various educational materials in the form of booklets, handbooks, textbooks, posters, activities, games, and practices were developed (UN/ISDR, 2006). "Disaster Reduction, Education and Youth" was the UN World Disaster Reduction Campaign in 2000, and in 2006–2007, there was the UN/ISDR campaign on "Disaster Risk Reduction Begins at School," both campaigns addressing the emphasis of integrating disaster risk reduction into education. The 2006–2007 "Disaster Risk Reduction Begins at School" campaign reaffirms the priority for action 3 of the HFA on use knowledge, innovation, and education to build a culture of safety and resilience at all levels (Table 1). UN/ISDR not only highlighted the importance of integrating disaster risk reduction into formal education, but at the same time emphasized the importance of community participation in order to achieve sustainability within the community (UN/ISDR, 2006). In addition to providing education, school buildings could also serve as temporary shelter for the community following disasters; thus, the safety of the buildings is important to ensure the safety of students as well as continuation of education following disasters, and the campaign also promotes safe construction and retrofitting of school buildings to withstand natural hazards (UN/ISDR, 2006).

Lessons learnt from the experiences include the following: (i) education is a process for effective disaster reduction; (ii) knowledge, perception, comprehension, and actions are the four important steps; (iii) schools and formal education play an important role in knowledge development; (iv) family-, community-, and self-education are important for comprehension of knowledge and implementing risk reduction actions; and (v) holistic education includes actions at the local level, as well as its policy integration (Shiwaku, 2007).

Responding to the call from the UN/ISDR 2006–2007 campaign, various international and/or regional conferences and workshops on school safety and school education were held, and countries have developed national action agenda addressing the issue on integrating disaster risk reduction into the education curriculum as well as ensuring safety of school buildings. The

Table 1. Key Activities of HFA Priority 3.

(i) ***Information management and exchange***

(a) Provide easily understandable information on disaster risks and protection options, especially to citizens in high-risk areas, to encourage and enable people to take action to reduce risks and build resilience. The information should incorporate relevant traditional and indigenous knowledge and culture heritage and be tailored to different target audiences, considering cultural and social factors.

(b) Strengthen networks among disaster experts, managers, and planners across sectors and between regions, and create or strengthen procedures for using available expertise when agencies and other important actors develop local risk reduction plans.

(c) Promote and improve dialogue and cooperation among scientific communities and practitioners working on DRR, and encourage partnerships among stakeholders, including those working on the socioeconomic dimensions of DRR.

(d) Promote the use, application and affordability of recent information, communication, and space-based technologies and related services, as well as earth observations, to support DRR, particularly for training and for the sharing and dissemination of information among different categories of users.

(e) In the medium term, develop local, national, regional, and international user-friendly directories, inventories, and national information-sharing systems and services for the exchange of information on good practices, cost-effective, and easy-to-use DRR technologies, and lessons learned on policies, plans, and measures for DRR.

(f) Institutions dealing with urban development should provide information to the public on disaster reduction options prior to constructions, land purchase, or land sale.

(g) Update and widely disseminate international standard terminology related to DRR, at least in all official United Nations languages, for use in program and institutional development, operations, research, training curricula, and public information programs.

(ii) ***Education and training***

(h) Promote the inclusion of DRR knowledge in relevant sections of school curricula at all levels and the use of other formal and informal channels to reach youth and children with information; promote the integration of DRR as an intrinsic element of the UN Decade of Education for Sustainable Development (2005–2015).

(i) Promote the implementation of local risk assessment and disaster preparedness programs in schools and institutions of higher education.

(j) Promote the implementation of programs and activities in schools for learning how to minimize the effects of hazards.

(k) Develop training and learning programs in DRR targeted at specific sectors (development planners, emergency managers, local government officials, etc.).

(l) Promote community-based training initiatives, considering the role of volunteers, as appropriate, to enhance local capacities to mitigate and cope with disasters.

(m) Ensure equal access to appropriate training and educational opportunities for women and vulnerable constituencies; promote gender and cultural sensitivity training as integral components of education and training for DRR.

(iii) ***Research***

(n) Develop improved methods for predictive multirisk assessments and socioeconomic cost–benefit analysis of risk reduction actions at all levels; incorporate these methods into decision-making processes at regional, national, and local levels.

Table 1. *(Continued)*

(o) Strengthen the technical and scientific capacity to develop and apply methodologies, studies, and models to assess vulnerabilities to and the impact of geological, weather, water and climate-related hazards, including the improvement of regional monitoring capacities and assessments.

(iv) ***Public awareness***

(p) Promote the engagement of the media in order to stimulate a culture of disaster resilience and strong community involvement in sustained public education campaigns and public consultations at all levels of society.

DRR: Disaster risk reduction.
Source: UN/ISDR (2007b).

International Conference on School Safety, held on January 18–20, 2007, in Ahmedabad, Gujarat, India, recognizes the importance of every child to receive education and live in a safe and sustainable environment and adopted the Ahmedabad Action Agenda for School Safety, which aims to achieve "Zero Mortality of Children in School from Preventable Disasters by the year 2015." To achieve its goal, the action agenda are outlined under immediate priority and long-term accomplishments (i.e., by 2015), with four priority areas, namely, (i) disaster education in school, (ii) disaster-resistant school infrastructure, (iii) safe school and community environments, and (iv) advocacy and government policy on school safety (Ahmedabad Action Agenda, 2007). In the same year in October, the Asia-Pacific Regional Workshop on School Education and Disaster Risk Reduction was held in Bangkok, Thailand, where 304 participants from 24 countries in the Asia Pacific region came together to discuss on ways to improve resilience of school communities struck by disasters or in hazard-prone areas. The output of the workshop was the adoption of Bangkok Action Agenda, which focuses on four priority areas, namely, (i) integrating disaster risk reduction into school education, (ii) strengthening disaster education for community resilience, (iii) making schools safer, and (iv) empowering children for disaster risk reduction (UN/ISDR, 2007c). The main focus of both agendas was on integrating disaster risk reduction into school education and ensuring the safety of school buildings, effectively complementing the core indicators of the HFA, particularly with the specific guidance on making schools disaster resilient and promoting participation of communities and children in the disaster risk reduction initiatives.

The Kashmir earthquake in 2005 affected over 3.5 million people, killed 87,350, and injured approximately 70,000 people in Pakistan. Over 17,000 school-age children died in the collapsed schools and over 20,000 were injured. Out of the 9,000 schools in the earthquake-affected regions in northern Pakistan, 8,000 were damaged beyond repair. In 2008, the International Conference on School Safety held in Islamabad issued the Islamabad Declaration on School Safety, which specified that "policies, guidelines, implementing and monitoring mechanisms are needed. This translates into actions that address identifying resilient school needs, retrofitting existing structures, creating evacuation plans and safe havens, improving community and student awareness through outreach and simulations. Selection of safe sites, design and construction technologies and materials also apply to the larger built environment" (Islamabad Declaration on School Safety, 2008). It also emphasizes the importance of safe school buildings to achieve disaster resilience in the education sector. The Declaration emphasized on the aspect of safe school buildings and seeks to achieve its aim through establishment of policies and partnerships between national government and local entities and community, reaffirming the call from the HFA on enhancing disaster resilience at schools. Action plans address school structural and nonstructural vulnerabilities and strongly encourages community participation. Community involvement is necessary since the community is the first responder to disaster situations and partnership will allow transfer of knowledge and practices, ensuring its continuation among the individuals as well as the community. Table 2 summarizes the three national initiatives described.

Table 2. Summary of the Three National Initiatives – Ahmedabad Action Agenda, Bangkok Action Agenda, and Islamabad Declaration on School Safety.

National Initiative	Priority Areas
Ahmedabad Action Agenda for School Safety	• Disaster education in school • Disaster resistant school infrastructure • Safe school and community environments • Advocacy and government policy on school safety
Bangkok Action Agenda	• Integrate disaster risk into school education • Strengthen disaster education for community resilience • Make schools safer • Empower children for disaster risk reduction
Islamabad Declaration on School Safety	• Identify school structural and non-structural vulnerabilities • Retrofit existing structures • Create evacuation plans and safe havens • Improve community and student awareness through outreach

In 2009, the UN launched the "One Million Safe Schools and Hospital" campaign to address and advocate the need to ensure that buildings, such as schools, are built with compliance to the safety standards to enhance disaster resilience. "People in unsafe schools, hospitals and health facilities are at the greatest risk of losing their lives," UNISDR said and continued, "Children in schools and the sick in hospitals and health facilities are the most vulnerable people in times of disaster" (UNC, 2010).

As discussed in the previous section, the importance of Education in Emergencies has also been advocated in recent years. The Inter-Agency Network for Education in Emergencies (INEE), an online network of educational stakeholders of nongovernmental organizations, UN agencies, donors, practitioners, researchers, and individuals, has also developed a similar document entitled *A Handbook of Minimum Standards for Education in Emergencies, Chronic Crises and Early Reconstruction* referred to as the "INEE Minimum Standards." Launched in 2004 at INEE's Second Global Inter-Agency Consultation on Education in Emergencies and Early Recovery in South Africa, the handbook is designed to provide governments and humanitarian workers the tools necessary to address the EFA and UN MDG (Inter-Agency Network for Education in Emergencies [INEE], 2002). The INEE Minimum Standards consists of five categories: (i) "Access and Learning Environment" focuses on partnerships to promote access to learning opportunities as well as intersectoral linkages; (ii) "Teaching and Learning" focuses on curriculum, training, instruction, and assessments, as the essential factors to promote effective teaching and learning; (iii) "Teachers and Other Education Personnel" focuses on the administration and management of human resources in the field of education, which includes recruitment and selection, conditions of service, as well as supervision and support; (iv) "Education Policy and Coordination" focuses on policy formulation, planning, implementation, and coordination; and (v) "Community Participation" and monitoring and evaluation are applicable to all the categories (INEE, 2002).

In an effort to promote the Education in Emergencies, the UNESCO International Institute for Education Planning (IIEP) developed the *Guidebook for Planning Education in Emergencies and Reconstruction*, consisting of five sections: (i) general overview (introduction, prevention of conflict and preparedness for disaster, challenges in emergencies and reconstruction, capacity-building, education for all in emergencies, and reconstruction); (ii) access and inclusion (rural populations, gender, ethnicity/political affiliation/religion, children with disabilities, former child soldiers, learning spaces and school, facilities, open and distance learning, technology, non-formal education, early childhood development, post-primary education); (iii) teachers and learners (identification, selection, recruitment of teachers, education workers, teacher motivation, compensation and working conditions, measuring and monitoring teachers' impact, teacher training: teaching and learning methods, psychosocial support to learners); (iv) curriculum and learning (curriculum content and review processes, health and hygiene education, HIV/AIDS preventive education, environmental education, landmine awareness, education for life skills peace, human rights and citizenship, vocational education and training, textbooks, educational materials and teaching aids); and (v) management capacity (assessment of needs and resources, planning processes, project management, legal frameworks, community participation, structure of the education system, data collection and education management information systems, budget and financial management, human resources: ministry officials, donor relations and funding mechanisms, coordination and communication) (UNESCO, 2010a).

The guidebook is primarily targeted at the staff of ministries of education – national, provincial, and district-level planners and managers – in disaster affected areas. The following are the three core aspects of education in emergencies emphasized by IIEP: (i) preparing the planning for emergency – an education sector diagnosis is necessary to examine the risk and vulnerabilities of the education system to both human and manmade disasters, which in turn help in developing budgeting for contingency planning, and capacity strengthening within the education system; (ii) responding to an emergency – including rapid assessment of education needs, coordination of education actors, and capacity gap analysis for reconstruction; and (iii) continuing education during and after an emergency – depending on the context, and strategies in ensuring continued provision of education such as temporary learning spaces as well as advocacy to donors, agencies, private sectors on funding education (UNESCO, 2010b).

As discussed in this section, nations have recognized the importance of integrating disaster risk reduction in education and safe buildings and have adopted national action agendas and declaration to promote school safety. However, is educational curriculum and safe building structures sufficient in achieving disaster resilience in schools? The next section addresses some issues faced in the process of implementing disaster risk reduction in education.

DISASTER EDUCATION POLICIES AND PROGRAMS

OECD Draft Policy

OECD is an international organization that conducts monitoring, analyzing, and forecasting, thereby providing public policy guidelines to help governments foster prosperity and fight poverty through economic growth and financial stability. It consists of 33 member countries, as well as international organizations (Organization for Economic Cooperation and Development [OECD], 2001). The 2009 *Draft Policy Handbook on Natural Hazard Awareness and Disaster Risk Reduction Education* suggests nine fundamental principles for disaster education as well as a strategy consisting seven focus areas (OECD, 2009). Among them some of the highlighted areas are as follows:

- Natural hazard awareness and disaster education efforts should aim at encouraging **voluntary** risk reduction activities.
- **Appropriate risk communication techniques** should be adopted to reach the targeted audiences and induce the desired changes in behavior and perception. **Specific and realistic measures for local conditions** are strongly recommended.
- **Continuous monitoring and periodic evaluations** of awareness and education efforts should be conducted to assure accountability and transparency to increase public confidence outcomes.

In addition, the OECD emphasized three crucial areas that require specific attention: gain scientific knowledge of hazards and risk hazard mapping (prioritization), identify the desired behavior and perception changes (awareness, public education), and identify roles and methods tools (stakeholder involvement and international cooperation).

The Pakistan National Education Policy of 2009

Pakistan, in particular the northern highlands, falls in the seismic active zone, is prone to earthquake as well other natural hazards such as floods, snow and ice avalanches, landslides, and river erosions. The Islamabad Declaration on School Safety was adopted at the International Conference on School Safety in Islamabad in 2008. In the following year, the Pakistan National Education Policy was issued (Ovington, 2010). The policy includes various aspects from strengthening education curricula and training to infrastructure and preventive measure. Curriculum, especially of social studies, geography, languages, and literacy, shall include themes on emergencies, natural disasters, and trauma management based on latest international best practices and shall also include information about response in an emergency or a disaster. A repository of all emergency-related materials, manuals, guidelines, minimum standards, and research pertaining to education shall be maintained at the teachers' training institutions, schools, college, and universities. Teacher education and training curricula shall include provisions to enable the teachers to address education in emergencies. Furthermore, Education in Emergencies (action 7) and rehabilitation measures (action 8) were mentioned as well.

Earthquake Safety Education Program, Iran

A successful disaster education program is the Earthquake Safety Education program in Iran. The program includes integration of disaster education into the formal education and performs non-formal education activities, simultaneously involving students at the local context as well as engaging participation from the community. Iran, being located near the faults, has high risks of seismic hazards and has suffered several earthquakes resulting in severe economic losses in the past. As such, the government has initiated the "School Earthquake Safety" initiative, "School Safety Act," as well as the "Earthquake Safety Education in School" (International Institute of Earthquake Engineering and Seismology [IIEES], 2006). The Earthquake Safety Education program is led by the Iranian Ministry of Education (MOE) in collaboration with IIEES as well as other public and civic organizations. Disaster lessons are integrated within science, geography, literature, and other curricula with preparedness and practice books designed for different grades and handbooks for teachers (Petal, 2008). Safety drills are one of the most commonly performed activities. Every year, the IIEES, Iranian MOE, National Committee for Natural Disaster Reduction, Iranian Red Crescent Society, Iran National Television and Radio, and other related national organizations organize these drills to prepare students and staff for appropriate and rapid responses during earthquakes. There is also a broadcast of an "Earthquake safety alarm" on national and local radio. Furthermore, a "School Earthquake Safety Council" involving parents and teachers voluntarily serves as a form of disaster risk reduction and preparedness effort at the individual school level (Petal, 2008). This case study shows the integration of disaster education into formal education, and at the same time, non-formal activities (i.e. safety drills) are conducted to increase preparedness and response among students. In addition, the program is not limited within school boundaries but reaches out to parents who voluntarily involve themselves in the "School Earthquake Safety Councils" with the teachers in their effort to achieve disaster risk reduction and preparedness.

Myanmar Education Recovery Program (MERP)

In 2010, the Ministry of Education, Union of Myanmar, together with UNESCO developed the MERP initiative, which seeks to enhance resilience in the education sector in Myanmar by addressing disaster risk reduction and emergency preparedness as an integral part of education, and integrates them in all the HFA's five priorities for action, through a community-based, participatory, and multisector approach. The initiative consists of a comprehensive training package Disaster Reduction in Education for township education officers, school principals, teachers, and students. The content of the training material addresses the five priorities for actions, as seen through an education lens, to help reduce risk, mitigate, and prevent the impacts in the education sector, namely, (i) developing institutional base for disaster risk reduction in education, (ii) identifying, assessing, and monitoring disaster risks in the education sector, (iii) building a culture of safety through disaster risk reduction in education, (iv) reducing the underlying risk factors in the education sector, and (v) preparing for effective emergency response and recovery in education (UNESCO, 2010c).

E-HFA: FUTURE PERSPECTIVE OF DISASTER EDUCATION POLICY

An integrated approach is necessary to ensure disaster risk reduction is incorporated into not only the schools but also into the education sector as a whole. The approach should not only consider education curricula and safe school buildings but also address legislative measures (i.e., having formal guidelines for implementation and funding), proper early warning systems and risk assessments, training of qualified professionals, promoting community involvement, as well as measures taken to prepare community in responding to disasters. The MERP initiative, which addresses HFA in the education sector, provides the basis for an integrated approach to disaster education. The document *Words into Action: A Guide for Implementing the Hyogo Framework*, by the UN/ISDR, provides advice on useful strategies for implementing the HFA. The guide describes 22 suggested tasks that are organized to help address and guide the implementation of the HFA's five priorities for action. Out of the 22 tasks, 16 were identified and modified to suit the education sector, and are considered as E-HFA (Education in Hyogo Framework for Action) (Table 3).

The 16 tasks identified should be performed at all levels (i.e., national, local, community/school) to achieve sustainable implementation. Table 4 summarizes suggested activities for the respective tasks. At the national and local levels, the Ministry of Education (MOE) and local education department would play key roles in performing the tasks. At the community level, it would be the principals, teachers, as well as students at schools.

Table 3. Proposed 16 Tasks Relevant to the Education Sector.

Priority 1: Developing institutional base for disaster risk reduction in education
1. Engage in multi-stakeholder dialogue to establish the foundation for disaster education
2. Create or strengthen mechanism for systematic coordination for disaster education
3. Assess and develop the institutional basis for disaster education
4. Prioritize disaster risk reduction and allocate appropriate resources for disaster education
Priority 2: Identifying, assessing, and monitoring disaster risks in the education sector
5. Establish risk assessments for the education sector
6. Strengthen early warning in the education sector through effective communication and dissemination mechanism.
Priority 3: Building a culture of safety through disaster education
7. Develop public program to raise awareness of disaster risk reduction
8. Include disaster risk reduction in the education system
9. Develop disaster risk reduction training and learning at community level
10. Enhance dissemination of disaster risk reduction information
Priority 4: Reducing the underlying risk factors in the education sector
11. Environment: Understand sustainable ecosystem, environmental and natural resources management
12. Establish measures to incorporate disaster risk reduction in urban and land-use planning
13. Structures: Strengthen mechanisms for improved building safety and protection of critical facilities in the education sector
14. Disaster recovery: Develop a recovery planning process that incorporates disaster risk reduction
Priority 5: Preparing for effective emergency response and recovery in education
15. Build on disaster preparedness capacities and mechanisms in the education sector
16. Assess disaster response preparedness capacities and mechanisms through strengthened planning

Table 4. Suggestions for Integrating Disaster Risk Reduction in the Education Sector – National, Local, and Community (i.e., school).

HFA Priorities for Action and their Respective Key Activities	National	Local	Community (i.e., schools)
HFA 1 • Multi-stakeholder dialogue • Systematic coordination	• Establish DRR multi-stakeholder group at all levels • Hold regular meetings • Establish national DRR education programs and ensure proper local implementation • Establish national disaster month to promote disaster risk reduction awareness		• Promote community involvement in the focal group
• Institutional basis	• Develop policy/Act promoting DRR integration into education curricula • Ensuring proper implementation and compliance to the policies and/or Acts		
• Allocation of resources	• Establish policy and/or regulations for specific amount of education budget to be allocated for DRR education	• Allocate specific amount of budget for DRR education	• Allocate specific amount of budget for DRR education
HFA 2 • Risk assessments	• Establish risk assessment standards on multi-hazards (i.e., earthquakes, typhoons, floods, and land subsidence) specific for schools • Establish evaluation system on the assessments on a regular basis	• Enforce risk assessment and evaluation system established by national government • Advocate risk assessment in the local community (ex. workshops) • Encourage community participation	• Encourage community and parental participation in risk assessment.

• Strengthen early warning through effective communication	• establish coordination between MOE and related bureaus and/or ministries • National/Local television and radio broadcast on DRR awareness information as well as disaster updates		• Internal early-warning via public address (PA) system • Coordination with local community with indigenous early warning systems • Use of disaster calendar
HFA 3			
• Public awareness programs • Incorporate DRR in education	• Develop national DRR education curricula • "Integrative" and/or "Curriculum Integration" approach may be preferred	• Advocate DRR awareness at local events • Conduct local activities such as community mapping	• Integrate DRR into school curricula. • Progressive learning is encouraged • Conduct non-formal DRR activities (ex. field-trips, town-watching, hazard map making etc..) to reinforce or expand upon the classroom curriculum • Encourage parental participation in the non-formal activities • Provide tasks that may be able to be conducted at home to evoke informal learning.
• Training and learning at community level • Dissemination of DRR information	• Promotional clips on national/local television and radio • Website • Conduct teachers' training (at national and local level). In addition to DRR information, training should also include emergency skills (i.e., first aid) and post-disaster psychological consultations Ensuring at least one teacher from all schools attend the training		
HFA 4 • DRR in land-use planning	• Establish land-use planning guidelines on multi-hazards (i.e., earthquakes, typhoons, floods and	• Enforce land-use planning and structural and non-structural safety guidelines established by national government	

Table 4. *(Continued)*

HFA Priorities for Action and their Respective Key Activities	National	Local	Community (i.e., schools)
• Strengthen building safety, protect facilities	land subsidence), specific for schools • Establish structural and non-structural safety guidelines on multi-hazards, specific for schools • Establish evaluation system on the assessments on a regular basis	• Advocate land-use planning and structural and non-structural safety guidelines (ex workshops) • Encourage community participation	
HFA 5			
• Disaster recovery planning	• Establish disaster recovery planning guidelines on multi-hazards (i.e., earthquakes, typhoons, floods and land subsidence), specific for schools	• Enforce disaster recovery planning established by national government • Advocate disaster recovery planning guidelines (ex workshops) • Encourage community participation	
• Enhance disaster preparedness capacities • Assess disaster response preparedness capacities and strengthen planning	• Establish disaster recovery planning guidelines on multi-hazards (i.e., earthquakes, typhoons, floods and land subsidence), specific for schools. • Specifications on evacuation shelter standards (assessing suitability of schools used as evacuation shelter)	• Enforce disaster recovery planning established by national government • Advocate disaster recovery planning guidelines (ex workshops) • Assessing schools to determine its suitability as evacuation shelter • Creating evacuation map • Establishing specific taskforce to perform specific tasks during emergency • Ensuring availability of emergency supplies locally	

DRR: Disaster risk reduction.

In HFA 1, the four key focuses are as follows: (i) multi-stakeholder dialogue, (ii) systematic coordination, (iii) institutional basis, and (iv) allocation of resources. With regard to HFA 1's focus areas (i) and (ii), establishment of a multi-stakeholder group at all levels is strongly recommended so as to develop a pool of disaster risk reduction specialists. Furthermore, regular meetings to discuss on disaster risk reduction issues in the education provide a platform for the members to be updated with the current progress and share experiences with each other. In terms of institutional basis, development of appropriate policy, Acts, and/or guidelines by the MOE to promote disaster risk reduction into the education curricula provides a standardized framework for implementation. Local education department plays a role in ensuring proper implantation and compliance to the policies, Acts, and/or guidelines. A feedback mechanism between the MOE and the local education department would be helpful for long-term monitoring of the effectiveness and efficiency of the implementation progress. In addition, the MOE may decide on a national disaster month and coordinate with the related departments in organizing various activities to promote awareness raising. The education department may also implement and organize activities in accordance with the local context to promote disaster risk reduction in the community during the "disaster month." Last but not least, it would be recommended that specific funds are allocated for the purpose of disaster education to ensure that disaster risk reduction funds are available.

HFA 2 focuses on risk assessments and early warning through effective communication. Risk assessment standards and evaluation system should be established by the MOE. Risk assessment should be limited to not only seismic capacity but also all hazards. The local education department and schools play a role in ensuring proper implementation and strict compliance to the standards. Furthermore, community participation in risk assessment should be encouraged. In terms of early warning systems, coordination between MOE and other related national departments as well as between the national and local governments is necessary. Coordination with the media (e.g., TV and radio) would be helpful. Within the school, PA system is necessary for emergency announcement. Practice of disaster calendar may help cultivate disaster preparedness among students.

In terms of HFA 3, which focuses on disaster education itself, the key areas are as follows (i) public awareness, (ii) incorporating disaster risk reduction into the education curricula, (iii) training, and (iv) dissemination of disaster risk reduction information. To achieve HFA 3, it is essential that MOE develops a set of disaster educational materials that can be integrated into the curricula. Integration may be through (i) curricula integration (i.e., specially developed units, modules, or chapters concentrating on disaster risk reduction), (ii) extracurricular integration (i.e., national and/or local campaigns), (iii) curriculum infusion (i.e., complete set of units focusing on disaster risk reduction), and (iv) stand-alone course (i.e., specialized course curricula focused on disaster risk reduction, which could be on a one-time basis) (Petal, 2008). Local education department and schools ensure proper integration and infusion based on the local context. At the local level, community mapping may be encouraged, which also serves as an opportunity for the community people to interact and exchange information. At the schools, activities such as town-watching, hazard mapping, and field-trips may help reinforce and/or expand the knowledge acquired from the classroom curriculum. Parental participation in school activities (e.g., evacuation drills, first-aid training) may also help enhance awareness among the community. Students may be asked to perform simple tasks at home (e.g., ensuring home furniture are well-anchored, preparing home emergency kit, etc.), which could serve as an opportunity for the students to interact with their parents and family members, and at the same time, spread the preparedness practice and disaster awareness within the family. In order that teachers are equipped with the knowledge and skills to teach, it is important for MOE to develop suitable disaster risk reduction training programs for the teachers. In addition, governmental officers (i.e., MOE and/or local education department), academicians, and/or practitioners (i.e., working on disaster risk reduction) may also attend such training courses.

HFA 4 focuses on land-use planning, building safety, and protection of facilities (Table 5). In addition, the priority for action also stresses on the need for disaster recovery planning (Patel, no date; WBDG, 2010; WMO, 2007). Similar to risk assessment in HFA 2, land-use planning should not be limited to seismic capacity but should also consider risks of other hazards such as typhoons, floods, and land subsidence. The standards for land-use planning, building safety, as well as structural and nonstructural safety would be determined by the MOE and enforced at local levels through the local education department and schools. Regular monitoring and evaluation is necessary and would be conducted by the local education department. Following disasters, class disruption poses a severe problem to students. Proper postdisaster recovery planning, which includes measures in ensuring class continuity, and disaster recovery actions are essential, and it is necessary for the MOE to provide proper guidelines for these purposes. Local education department and schools enforce these guidelines and may modify them according to the local context.

Table 5. Suggestions on Land-Use Planning, Building, Structural and Nonstructural Safety for Typhoons, Flood and Land Subsidence Hazards.

	Typhoon	Flood	Land Subsidence
Land-use planning	• Sediment load • Soil characteristics ex water storage capability, infiltration rates, type and density of vegetative cover • Geomorphology and connectivity with river channel and flood plains		
Building safety	• Structure design that withstand lateral and uplift forces • Doors and windows protected by covering and/or bracing	• Flood prone locations should be elevated to reduce chances of flooding • Making building water tight to prevent water entry • Making parts of the building resistant to water • Building of levees and/or floodwalls if necessary	• To enhance subsidence-resistant, shear walls, geofabrics and earth reinforcement techniques such as dynamic compaction may help to increase resistance to subsidence and stabilize collapsible soils.
Structural and nonstructural safety	• Structural elements: location, shape, number of stories, roof type, material of construction, alterations in original design • Nonstructural elements: stairways, doors, windows, lighting fixtures, pipes, wall cladding, false ceiling, location of equipments, anchorage, material of anchorage, location of hazardous material storage		

Last but not least, preparing for effective emergency response and recovery in education is important, as mentioned in HFA 5. The emphasis in this priority for action would be on the following: (i) enhancing disaster preparedness capacities and (ii) assessing disaster response preparedness capacities. To achieve this, proper pre-disaster recovery planning and assessing the preparedness capacities are essential, which in turn helps prepare appropriate response capacities. Areas to be considered would include establishing standards on evacuation shelters so as to assess suitability of schools being used as an evacuation shelter as well as establishing guidelines on setting up specific task forces and their respective responsibilities as well as feedback mechanisms during emergencies. Guidelines and standards should be developed by the MOE and enforced by the local education department and schools. Within the community and schools, creating a local evacuation map may serve to enhance preparedness and response among the community people and students.

When considering disaster education, it should not only be limited to the education curricula, but should also include related issues such as structural and nonstructural safety; legislative measures supporting the integration, implementation, as well as sufficient funding; proper early warning systems and risk assessments; training of qualified professionals, etc. Therefore, an integrated approach is necessary, and the suggestions made seek to address the various issues so as to achieve the aim of enhancing disaster resilience in the education sector.

ACKNOWLEDGMENTS

The first author would like to thank the support from ASJA International Scholarship.

REFERENCES

Ahmedabad Action Agenda for School Safety. International Conference on School Safety, 18–20 January (2007). Ahmedabad, Gujarat, India. Available at http://www.ineesite.org/toolkit/docs/doc_1_87_Agenda_for_School_Safety.pdf (Accessed on May 07, 2010).

Inter-Agency Network for Education in Emergencies (INEE). (2002). INEE Minimum Standards. Available at http://www.ineesite.org/ (Accessed on February 15, 2011).

International Institute of Earthquake Engineering and Seismology (IIEES). (2006). *Case study: IRAN. School disaster risk reduction: Think globally, act locally*. Available at http://www.iiees.ac.ir/iiees/English/Publicedu/school_safety_iran_case_study_davos.pdf (Accessed on November 20, 2010).

Islamabad Declaration on School Safety. International conference on school safety Islamabad, Pakistan, 14–16 May 2008.

Organization for Economic Cooperation and Development (OECD). (2001). Available at http://www.oecd.org/ (Accessed on December 9, 2010).

Organization for Economic Cooperation and Development (OECD). (2009). *Draft policy handbook on natural hazard awareness and disaster risk reduction education.* Available at http://www.oecd.orgdataoecd15043730809.pdf (Accessed on November 30, 2010).

Ovington G. (2010). DRR in education. In: DRR High Level Experts' Group Meeting, November 10, 2010, Islamabad, Pakistan. UNICEF, UNESCAP. Available at http://www.unescap.org/idd/events/2010_Pakistan_Floods_II/Unicef_Education.pdf

Patel T. (no date). Risk and vulnerability assessment of health facilities: Focus on structural and non-structural elements. Available at http://www.preventionweb.net/files/11214_Riskandvulnerabilityassessmentofhea.pdf (Accessed in December 2010).

Petal M. (2008). *Concept note: Formal and informal education for disaster risk reduction.* Available at www.riskred.org/activities/ddredislamabad.pdf (Accessed on November 20, 2010).

Shiwaku K. (2007). Towards innovation in school disaster education: Case research in Kathmandu, Nepal, PhD Thesis, Kyoto University.

UN. (2010). *Millennium development goals.* Available at http://www.un.org/millenniumgoals/ (Accessed on November 19, 2010).

UN News Centre (UNC). (2010). *UN seeks to improve safety of schools and hospitals in natural disasters.* Available at http://www.un.org/apps/news/story.asp?NewsID=34321&Cr=quake&Cr1=(Accessed on November 25, 2010).

UNEP. (1992). *Agenda 21/Chapter 36: Promoting education, public awareness and training.* Available at http://habitat.igc.org/agenda21/a21-36.htm (Accessed on January 12, 2010).

UNESCO. (2005a). *Education for sustainable development. Strategy.* Available at http://www.unesco.org/en/esd/strategy/ (Accessed on December 9, 2010).

UNESCO. (2005b). *UN Decade of education for sustainable development.* Available at http://www.unesco.org/en/education-for-sustainable-development/decade-of-esd/ (Accessed on November 10, 2010).

UNESCO. (2010a). International institute for education planning, *Guidebook for planning education in emergencies and reconstruction.* Available at http://www.iiep.unesco.org/fileadmin/user_upload/Cap_Dev_Technical_Assistance/pdf/Guidebook/Guideboook.pdf (Accessed on January 20, 2011).

UNESCO. (2010b). International institute for education planning, education in emergencies. Available at http://www.iiep.unesco.org/en/focus-on-education-in-emergencies/education-in-emergencies-an-introduction.html (Accessed on January 20, 2011).

UNESCO. (2010c). *Myanmar education recovery programme (MERP). Building the Resilience in the Education Sector to Disasters.* Available at http://www.unescobkk.org/en/education/education-units/esd/projects/post-conflict-and-post-disaster-responses/unesco-asia-pacific-initiatives/myanmar-education-recovery-project/ (Accessed on May 1, 2010).

UN/ISDR. (2005). *Thematic cluster/platform on knowledge and education.* Available at http://www.unisdr.org/eng/task%20force/working%20groups/knowledge-education/knowledge-education.htm (Accessed on November 10, 2009).

UN/ISDR. (2006). *World Disaster Reduction Campaign. Disaster Risk Reduction Begins at School.* Available at http://www.unisdr.org/eng/public_aware/world_camp/2006-2007/pdf/WDRC-2006-2007-English-fullversion.pdf (Accessed on January 28, 2010).

UN/ISDR. (2007a). Regional Analysis on Disaster risk reduction education in the Asia-Pacific Region – in the context of Hyogo Framework for Action Priority 3 Implementation. 2nd Asian Ministerial Conference on Disaster Risk Reduction, 7–8th November 2007, New Delhi,

India. Available at http://nidm.gov.in/amcdrr/acdrrbro.pdf (Accessed on November 29, 2010).

UN/ISDR. (2007b). *Hyogo Framework for Action 2005-2015: Building the Resilience of Nations and Communities to Disasters*. Available at http://www.unisdr.org/eng/hfa/docs/= Hyogo-framework-for-action-english.pdf (Accessed on January 28, 2010).

UN/ISDR. (2007c). *Bangkok Action Agenda*. Outcome of the Asia-Pacific Regional Workshop on School Education and Disaster Risk Reduction, 8–10 October 2007. Available at www.unisdr.org/asiapacific/ap-whatnew/Regional-thematic-report-education-agenda.pdf (Accessed on January 25, 2010).

UN/ISDR. (2008). Regional analysis on disaster risk reduction education in the Asia-Pacific Region – in the context of Hyogo Framework for Action Priority 3 Implementation. 3rd Asian Ministerial Conference on Disaster Risk Reduction, 2–4 December 2008, Malaysia. Available at www.preventionweb.net/.../3089_3089FirstAnnouncementAsian-MinisterialConference.pdf (Accessed on November 29, 2010).

Whole building design guide (WBDG). (2010). Resist natural hazards. Available at http://www.wbdg.org/design/resist_hazards.php (Accessed on December 20, 2010).

Wisner, B. (2006). Let Our Children Teach Us! A Review of the Role of Education and Knowledge in Disaster Risk Reduction. Inter-Agency Task Force Cluster Group on Education and Knowledge. Available at http://scholar.google.co.jp/scholar?q = 2006 + Review + of + the + Role + of + Education + and + Knowledge + in + Disaster + Risk + Reduction&hl = ja&as_sdt = 0&as_vis = 1&oi = scholart (Accessed on November 30, 2010).

World Meterological Organization (WMO). (2007). Formulating a basin flood management plan – a tool for integrated flood management. Available at http://www.apfm.info/pdf/ifm_tools/Tools_Basin_Flood_Management_Plan.pdf (Accessed on December 20, 2010).

CHAPTER 3

ROLES OF SCHOOL IN DISASTER EDUCATION

Koichi Shiwaku and Glenn Fernandez

INTRODUCTION

In addressing disaster management in schools, many researchers and workers in NGOs, UN agencies, and other organizations have pointed out that school-building safety and disaster education are significant factors in developing school safety, especially in the case of earthquake disasters (Izadkhan, 2004; Dixit, 2004; Wisner et al., 2004). School-building safety is useful for disaster reduction in the short term, while disaster education can play a significant role in developing a culture of disaster reduction in the long term. The importance of disaster education at the school level is recognized in the works of Radu (1993), Kuroiwa (1993), Arya (1993), Frew (2002), and Shaw, Shiwaku, Kobayashi, and Kobayashi (2004). Students are viewed as initiates into tradition, and parents are also congregational members (Strike, 2000). Shaw and Kobayashi (2001) stress that schools play an important role in raising awareness among students, teachers, and parents. UNISDR conducted a campaign based on the observation that children are among the most vulnerable population group during disasters (UNISDR, 2007a) and that disaster risk education empowers children and helps build greater awareness of the issue in communities (UNISDR, 2007b).

School education hours can be divided into curricular and extra-curricular. In many parts of the world, there is no separate or special curriculum for

Disaster Education
Community, Environment and Disaster Risk Management, Volume 7, 45–75

ISSN: 2040-7262/doi:10.1108/S2040-7262(2011)0000007009

disaster education in elementary and high schools. Therefore, school disaster education should be conducted as part of both curricular and extra-curricular activities in order to increase the hours spent on disaster education and to sustain disaster education. In many cases, extra-curricular activities can provide more hours to disaster education, compared to curricular activities. This gives the advantage of providing active learning to students. This chapter is divided into three parts. The first part discusses the Environment and Disaster Mitigation Course established in a high school in Japan. It is an example of a school disaster education curriculum. The second part is about the linkage between disaster education and the current curriculum in the case of Nepal. The third part is about disaster education in extra-curricular activities, featuring the science club activities in the Philippines.

ENVIRONMENT AND DISASTER MITIGATION COURSE IN MAIKO HIGH SCHOOL, JAPAN

Maiko High School, one of the public high schools in the Hyogo Prefecture of Japan, established its "Environment and Disaster Mitigation Course" in 2002. This course is the first of its kind at the high school level in the world, which focuses on disaster management. According to Maiko High School, the goal of disaster education is to equip citizens with disaster mitigation literacy (Maiko High School, 2002). Disaster mitigation literacy consists of three factors: fundamental knowledge and fundamental skills to cope with disasters and a strong will to contribute to society. The purpose of the course is to have students think of how we live and exist in a symbiotic society by utilizing the lessons learned from the Great Hanshin-Awaji Earthquake disaster. The course gives the students the disaster mitigation education concerning both the natural environment and the social environment. The fundamental concepts of the disaster mitigation education are the following:

1. The disaster mitigation education is based on the lessons from the Great Hanshin-Awaji Earthquake. It also makes the students think of the importance of life, cultivates the students' power against disasters, and brings up human beings who can contribute to the society.
2. The students are expected to understand deeply the various environments (the natural environment and the social environment) by learning the mechanism of natural phenomena and the relationship between disasters and human society.

3. Maiko High School cooperates with universities, research institutes, and relevant organizations. The students' understanding of the environment and disaster mitigation is deepened through experiential learning. To raise the students' attitude to "Think Globally, Act Locally" is one of the main goals to be attained. Maiko High School aims to produce individuals who can take actions independently.

The abilities that the course required students to develop are fundamental knowledge, comprehensive ability, and interest.

Curriculum

Table 1 shows the education curriculum of the "Environment and Disaster Mitigation Course" in 2002. The credit means the number of school hours that are used in the subjects per week. The subjects highlighted in gray are the newly established disaster education subjects. Roughly one-third of the units are used for the special subjects. The following are the brief explanation of each of the special subjects (Shiwaku, 2004):

- Disasters and Human Beings I: By studying the Great Hanshin-Awaji Earthquake, students understand the importance of lives, relationship among people, and cooperation. Students also understand the concept of environment and disaster management, relationship between the natural and social environment, and disaster management. Lectures by visiting lecturers, studies outside the classroom, reporting, presentation, discussion, and other practical activities are done.
- Environment and Science: Students learn about disaster management from the aspect of the natural environment. They learn not only the Great Hanshin-Awaji Earthquake but also local flood disasters. In addition, a field trip to a local mountain and visit to the museum of nature are conducted.
- Computer A: The ability to solve problems through information collection in the theme of disaster management is enhanced. Students get information literacy and the ability to utilize information networks.
- Natural Environment and Disaster Mitigation I: Students learn chemistry to understand environmental problems caused by human activities. Through experiments and observations, students think of the causes of environmental problems, their current status, and mechanism.
- Social Environment and Disaster Mitigation I: Students learn the link between social environment and disaster management. They learn the

Table 1. Education Curriculum of the Environment and Disaster Mitigation Course.

<table>
<tr><th>Credit</th><th>1st Year</th><th colspan="3">2nd Year</th><th colspan="3">3rd Year</th></tr>
<tr><td>1</td><td rowspan="4">Japanese 1</td><td rowspan="3" colspan="3">Japanese 2</td><td rowspan="3" colspan="3">Japanese 3</td></tr>
<tr><td>2</td></tr>
<tr><td>3</td></tr>
<tr><td>4</td><td rowspan="2">Geography A</td><td rowspan="2" colspan="2">World History B</td><td rowspan="2">World History A</td><td rowspan="2" colspan="2">World History B</td></tr>
<tr><td>5</td><td rowspan="2">Contemporary Society</td></tr>
<tr><td>6</td><td rowspan="4">Mathematics 2</td><td rowspan="4" colspan="2">Choose two subjects
Mathematics 2
Japanese Classics 1
Reading</td><td rowspan="3">Mathematics 3</td><td rowspan="3">Mathematics 2</td><td rowspan="3">Japanese Classics 1</td></tr>
<tr><td>7</td><td rowspan="4">Mathematics 1</td></tr>
<tr><td>8</td></tr>
<tr><td>9</td><td rowspan="2">Mathematics B</td><td rowspan="2" colspan="2">Geography A</td></tr>
<tr><td>10</td><td rowspan="2">Chemistry 1B</td><td rowspan="2" colspan="2">Chemistry 1A</td></tr>
<tr><td>11</td><td>Health Education</td><td rowspan="2">Chemistry 1B</td><td rowspan="2">Reading</td><td rowspan="2">Workshop</td></tr>
<tr><td>12</td><td rowspan="2">Physical Education</td><td rowspan="2">Physics 1B</td><td rowspan="2">Biology 1B</td><td rowspan="2">Biology 1A</td></tr>
<tr><td>13</td><td rowspan="2">Physics 1B</td><td rowspan="2">Biology 1B</td><td rowspan="2">Computer B</td></tr>
<tr><td>14</td><td rowspan="2">Domestic Technology</td><td rowspan="2" colspan="3">Physical Education</td></tr>
<tr><td>15</td><td rowspan="2">Physics 2</td><td rowspan="2">Japanese Classics</td><td rowspan="2">Choices α</td></tr>
<tr><td>16</td><td rowspan="4">English</td><td rowspan="3" colspan="3">Music/Art/Japanese Calligraphy</td></tr>
<tr><td>17</td><td rowspan="3" colspan="3">Physical Education</td></tr>
<tr><td>18</td></tr>
<tr><td>19</td><td rowspan="3" colspan="3">English</td></tr>
<tr><td>20</td><td rowspan="2">Computer A</td><td rowspan="4" colspan="3">Reading</td></tr>
<tr><td>21</td></tr>
<tr><td>22</td><td rowspan="4">Environment and Science</td><td rowspan="2" colspan="3">Human Beings & Society, Social, Welfare, Volunteerism, and Mental</td></tr>
<tr><td>23</td></tr>
<tr><td>24</td><td rowspan="2" colspan="3">Disasters & Human Beings</td><td rowspan="2" colspan="3">Activity in Disaster Mitigation II</td></tr>
<tr><td>25</td></tr>
<tr><td>26</td><td rowspan="5">Disasters & Human Beings I</td><td rowspan="2">Natural Environment & DM I</td><td rowspan="2" colspan="2">Social Environment & DM II</td><td rowspan="2">Natural Environment & DM II</td><td rowspan="2" colspan="2">Social Environment & DM II</td></tr>
<tr><td>27</td></tr>
<tr><td>28</td><td rowspan="2" colspan="3">Activity in Disaster Mitigation I</td><td rowspan="2" colspan="3">Graduation Report</td></tr>
<tr><td>29</td></tr>
<tr><td>30</td><td colspan="3">Long Home Room</td><td colspan="3">Long Home Room</td></tr>
<tr><td>31</td><td>Long Home Room</td></tr>
</table>

roles of the stakeholders in the disaster management cycle. Visiting lecturers are also invited.

- Activity in Disaster Mitigation I: One of the activities is the international exchange with Nepal. Students learn about the importance of local knowledge in a developing country. In other activities, students learn about reconstruction by visiting one of the areas most affected by the Great Hanshin-Awaji Earthquake. The learning style employed is finding and solving problems through preliminary learning, town watching, reporting, and presentation.
- Human Beings and Society, Social Welfare, Volunteerism, and Mental Health: Students think about how they should live in society. Social

welfare, volunteer activities, and mental health are the main learning topics.

- Disasters and Human Beings II: Students learn about risk management at various levels covering local and central governments, countries, the United Nations, and others.
- Graduation Report: Students select their preferred topic related to environment and disaster management and study it. The outputs are a thesis, PowerPoint presentation file, video, book, model, game, and various other products of the learning process.
- Activity in Disaster Mitigation II: The aim is to promote action-based learning. Students learn about recent disasters and read English articles. International and domestic exchange is promoted.
- Natural Environment and Disaster Mitigation II: Students gain knowledge of chemicals. The ability to analyze data by themselves is enhanced.
- Social Environment and Disaster Mitigation II: Students learn integrated disaster management. Not only natural disasters but also wars, accidents, and other human disasters are studied in order to achieve a safe and comfortable society.
- Computer B: Using information tools and equipment are emphasized. The ability to disseminate information on the Web is enhanced. Students also try to use GIS and learn its utilization for disaster management.
- Workshop: It is a kind of group work. Students make a manual or textbook of disaster management through their discussions. The ability to become disaster management leaders is enhanced.

The subjects mentioned above are those in 2002. "Workshop" and "Disasters and Human Beings II" are no longer offered now, and "Reading on Environment and Disaster Mitigation" was added as new subject. This modification was done to meet the actual conditions and needs of the students. In case of the first-grade students, "Computer A" and "Environment and Science" are advanced subjects for disaster management. Even in the general course, computer or science subjects are taken up. The "Environment and Disaster Mitigation Course" modified such subjects into special subjects. "Disasters and Human Beings I" is the subject with a new concept that cannot be covered by the regular education curriculum. It is a distinguishing subject of the course. Therefore, the subjects in the course are divided into three types:

1. General subjects
2. Special subjects integrated into general subjects
3. Special subject with a new education concept

Examples of Disaster Education Programs

The "Environment and Disaster Mitigation Course" provides a new kind of education to students because disaster management is not systematically covered in the regular education curriculum in junior high school and elementary school. As mentioned before, some special subjects have a new education concept. In the first grade, "Disasters and Human Beings I" is the newly established subject with characteristic features. This section shows the characteristic education programs in the subject and other education activities outside the school.

The education programs are classified as follows:

1. Lectures by Maiko High School teachers
2. Lectures by visiting lecturers (practitioners, university professors, researchers)
3. Activity-based learning inside schools
4. Activity-based learning outside schools
5. Practice

According to Shiwaku and Shaw (2008) and Maiko High School (2010), the following are the explanation of each of the characteristic education programs:

- Lecture on the general concepts of the "Environment and Disaster Mitigation Course": The Maiko High school teachers describe the contents of the course and show examples of the relationship between the natural environment and disasters, and its implication in the Great Hanshin-Awaji Earthquake. Students need to write a report on the lecture to get credit.
- Lecture by the Kansai Electric Power Co., Inc.: This describes the recovery work after the Great Hanshin-Awaji Earthquake and the lessons learned on lifeline risk management. Students need to write a report on the lecture to get credit.
- Lecture by the water works department of Kobe City: Recovery work after the Great Hanshin-Awaji Earthquake and the current measures against disasters are described. Students need to write a report on the lecture to get credit.
- Lecture by a professor of Kobe University: This topic describes the reasons why disaster education is important by showing actual examples. Students need to write a report on the lecture to get credit.
- Lecture by a professor of Kyoto University: The importance of emphasizing prevention and mitigation and enhancing social capacity to

disasters is described. Students need to write a report on the lecture to get credit.

- Wall newspaper on the Kobe Earthquake: Students are divided into 10 groups. Teachers give each group a theme related to the Great Hanshin-Awaji Earthquake, for example, "Fire department," "Railway Transport," "Electricity," "Gas," "Water Service," "Evacuation Center," "Volunteer Activity," "Mental Health Care," etc. Each group investigates its own theme, creates a poster, and makes a presentation (see Fig. 1).
- Writing a paper on the flood risk of Kobe: Teachers give students the theme "the Great Hanshin Flood." Each student makes some literature survey and prepares a paper on their preferred topic concerning "the Great Hanshin Flood." Completed papers are attached to the wall of the classroom, and mutual evaluation by students is also accomplished.
- Panel discussion: A panel discussion is done using different themes: "Being part of government, being a volunteer, and being a citizen," "the lessons of the Great Hanshin-Awaji Earthquake," and others.
- Providing lectures in elementary schools: As part of proactive learning, the students provide lectures in elementary schools, demonstrating that large stones and large trees flow as part of debris in mud flow or during a cyclone and may cause significant damages to the inhabitants. A video of the Great Hanshin Flood of 1938 is shown (see Fig. 2).

Fig. 1. Discussion among Students during Group Work (photo taken by author).

Fig. 2. Lecture to Elementary School Students (photo taken by author).

- Making a safety map with elementary school students: The students of the course and students in an elementary school are distributed in mixed groups. Each group conducts town watching. During the town watching, the students of the course talk about disaster management with the elementary school students. As output, each group makes a safety map and gives a presentation on the map.
- Visiting the Nojima Fault Conservation Center: Experiences of the Great Hanshin-Awaji Earthquake is transferred by a person who has the role to disseminate his experiences. The lecture is named as "Nojima Fault and the Mechanism of Earthquakes." The lecture is complemented with a visit to the Nojima Fault Museum. Each student hands in a report on this lecture.
- Fieldwork on Mount Rokko: Following the lectures of the university professors, a fieldwork to Mount Rokko is arranged. Mount Rokko is located in the northern part of Kobe City. The main purpose of the lecture is to understand the linkage of the natural and built environments. Each student submits a report on this lecture.
- Experience in the school for fire fighting: The students have a first-hand experience in the school in fire fighting in Kobe for two days. They experience training in different disciplines: training in rescue from the rubble, preparing meals outdoor, training in information collection, sleeping in the gymnasium, training for floods, training for tail water, training in using the fire extinguisher, training in connecting loops, and

getting introduced to the activities of the community for disasters, the welfare of the school, and the measures taken during the disasters in Kobe.

- 1.17 Memorial Event: 1.17 means January 17, which was the date of the Great Hanshin-Awaji Earthquake. The event is held for all students including those in the general course of the school. The local government, the Japan Self-Defense Force, the community, and the parents are also invited. Students of the Environment and Disaster Mitigation Course prepare for the event in the special subject and organize the event. The event provides students with practical work.
- Volunteer activities in affected areas: Volunteer activities are of two types. One involves the students helping the affected area from outside. Fund-raising and sending messages to the affected areas belong to this type of activity. The other type involves the students visiting the affected area to help people there. These voluntary activities are also examples of practical work.

Characteristic Points of the Course

Following are the characteristics of the education provided by the Environment and Disaster Mitigation Course at Maiko High School:

- Proactive learning (co-learning): The course emphasizes that students consider disaster management by themselves. The characteristic activity process involves establishing point issues, investigating, participating in group discussions, drawing conclusions, making presentations, and conducting evaluations. This process can make students notice future challenges or tasks and motivate themselves. This process can achieve the aim of making students learn by themselves even after their graduation from high school.
- Natural and social environment: A natural hazard is an event, but a disaster is related to human and social issues. It is important for disaster education to include social issues as well as natural environment issues. The course emphasizes social environment issues including welfare, lifeline, government, and others.
- Knowledge and awareness (motivation), and theory and practice: Knowledge is important in order for the students to be able to act properly. But actual action is more important to reduce the impacts of disasters. In this regard, awareness is important. The course invites visiting lecturers to provide knowledge or theory and requests students to do practical exercises through many programs in order to raise their awareness.

- Education for citizens: Generally, disaster education can be considered as education to save lives and help others. Making safer and more comfortable communities is important in both daily life and disaster situation. For example, helping vulnerable people in daily life is welfare and helping them in disaster education is disaster management. Safer community in daily life leads to safer community in disaster situation. Therefore, education for citizens is a significant aspect of disaster education.

Toward the Promotion of the Lessons Obtained from the Course

This section proposes many lessons on disaster education as the characteristic points. To promote the lessons of the course, not only schools but also other institutes or stakeholders need to make an effort. Local government generally emphasizes teacher's training and development of textbook or education materials. Both are important. But disaster education should meet local context. A standard program/curriculum is effective to some extent, but capacity building of teachers including management ability should be enhanced in order to make education more effective (Shiwaku, Shaw, Kandel, Surya, & Dixit, 2006). The local government is hoped to have the principle that schools develop their own curriculum for disaster education through implementation and that the process can be a part of teacher's training. To promote disaster education with the same concept as the Environment and Disaster Mitigation Course, schools should utilize both curricular and extra-curricular education. Schools should explore how disaster education can be integrated into the existing regular school curriculum. As for extra-curricular disaster education, practical learning should be emphasized. Schools need to motivate teachers to start disaster education because school teachers can be change agents who can promote school-based disaster education. The school is a key institution for promoting disaster education, but it is difficult for the school to have special knowledge on disaster management. In this regard, NGOs and research institutes play an important role. For example, the Environment and Disaster Mitigation Course also utilized special organizations or institutions to provide special knowledge to Maiko High School students through lectures. In addition, community-based organizations are also useful to bridge schools and special organizations and institutions in order to communicate school needs to such organizations. To promote school disaster education, related organizations/institutions should also be involved.

LINK BETWEEN SCHOOL CURRICULUM AND DISASTER EDUCATION IN NEPAL

Introduction

It is important to provide disaster education in school continuously. In this regard, curricular education is significant to be utilized for disaster education. This section considers the link between school curriculum and disaster education in Nepal as an example.

In Nepal, like most other places, school education is divided into curricular and extra-curricular education. The curricular education is the set of the fundamental subjects like mathematics, science, and social studies, which are decided by the government. In extra-curricular education, the school provides some activities that students can participate in outside of their regular school time, like cultural programs. Extra-curricular education is dependent on the school principal, and schools can organize programs as per their own wish. For disaster education in Nepal, no official special subject or program is established. One of the advantages of this is that the same education programs or contents can be provided to all students across the country. This is made possible when all schools follow a uniform standard curriculum.

After the 10th grade, which is the final grade of secondary education, students have to take an examination called "School Leaving Certificate (SLC)" to prove that they have sufficient knowledge and abilities as 10th-grade students. Only after qualifying the SLC can they be permitted to enter higher secondary level. At this level, students select one of three specialized courses: science, humanity, and business. Hence, the secondary level (9th–10th grade) can be considered as the final education level in which students can take the same type of education in school together. This section identifies linkage between disaster education and curriculum through analysis of the education of 9th and 10th grade as an example.

Disaster-Related Topics in the School Curriculum

Shiwaku (2007) identified some necessary contents of disaster education, which are as follows:

- Causes and mechanisms of disasters
- Disaster management cycle

- Experiences, problems, measures, and lessons in each stage of the disaster management cycle
- Roles of the government, communities, individuals, and other agencies in each stage of disaster management
- Natural environment and social environment

In an interview with an NGO in Nepal, it was concluded that science, environment, and social studies are the subjects more related to disaster management, compared to other subjects. Therefore, this section focuses on these three subjects. These subjects are reviewed by looking at textbooks and the tendency of the subjects to be related to disaster management is considered in terms of direct and indirect topics. Direct topics are topics that are obviously related to disasters or disaster management. Indirect topics are topics that are not apparently related to disasters or disaster management but can be covered as part of the disaster education contents mentioned above. For example, the natural and social environments are important factors that should be covered in disaster education (Shiwaku, 2004). But it is sometimes difficult for teachers to understand these environments and disaster management if the link between disaster management and these environments are not clear. Therefore, some of these kinds of topics are regarded as indirect topics in this section. Both direct and indirect topics are classified into the following:

- Mitigation
- Preparedness
- Response
- Recovery/reconstruction
- Cause of disaster (hazard)
- Natural environment
- Livelihood

Table 2 shows the direct topics in science in the 9th and 10th grades according to the textbooks used (Curriculum Development Center, 2005a, 2005b). Causes, effects, and mitigation are the main direct topics. Mitigation measures for floods and landslides are mentioned in the textbooks, but it is not clearly shown who should have responsibility to take the necessary measures. For the effects of disasters, the textbooks show a small range of damages, like construction damages. By learning the direct topics, students can acquire knowledge only on what disasters are (their mechanism), and it is difficult for them to establish the link between disasters and livelihood.

Table 3 shows the indirect topics in science in the 9th and 10th grades. Mitigation measures, problems (effects), causes, and the natural

Table 2. Topics in the Science Subject Directly Related to Disaster Management.

Mitigation	• Mitigation measure of flood • Mitigation measures of landslide (plants, terrace cultivation system, avoid overgrazing)
Preparedness	
Response	• Effects of floods to lives, property and man-made structure • Effects of landslides to houses and people • Effects of earthquakes (effects to human constructions, effects to life lines, diseases caused by pollution because of earthquakes)
Recovery/ Reconstruction	• Effects of volcanic eruptions (require a huge amount of money for rehabilitation)
Cause	• Mechanism of floods • Rain and flood • Causes and mechanism of landslides • Tree, rain, human activities, agriculture in sloping land • Causes and mechanism of earthquakes (earthquake and volcano, tectonic forces, expansion and contraction) • Causes and mechanism of volcanic eruptions
Natural Environment	• Effects of flood (flood deposits sand, soil, stone; effects to land and crops, effects to forests, soil erosion) • Effects of landslides (landslide carries away cultivable land, soil erosion) • Mitigation measures for landslides (plants and terrace cultivation system) • Effects of earthquakes (effects to land and weak points, effects to rivers, effects to life lines, diseases caused by pollution because of earthquakes) • Effects of volcanic eruptions (effects to the environment, effects to fertile land)
Livelihood	• Effects of floods to crops • Tree, rain, human activities, agriculture in sloping land • Mitigation measures for landslides (avoid overgrazing)

environment are the main topics. But the linkage between these topics and disaster management is not mentioned. If teachers have enough knowledge, then they can explain greenhouse effect and climate change. And then they can show that climate change has the ability to cause flood or drought. However, such description is not in the textbooks. The number of indirect topics is the same as the direct topics. If teachers can include the indirect topics as part of disaster education, the number of disaster education topics becomes doubled without any development of a new curriculum. Table 4

Table 3. Topics in the Science Subject Indirectly Related to Disaster Management.

Mitigation	• Mitigation measures for ozone layer protection
Preparedness	
Response	• Effects of chemical pollution to the people in Japan • Effects of common colds, mumps, measles, polio, rabies, AIDS • Effects of ozone layer depletion (human health, plants, temperature increase)
Recovery/Reconstruction	
Cause	• Causes of chemical pollution • Causes of common colds, mumps, measles, polio, rabies, AIDS • Causes and mechanism of ozone layer depletion • Causes and mechanism of greenhouse effect (temperature, water cycle, polar ice) • Causes and mechanism of pollution caused by industrial gas (diseases, air pollution, water pollution, acid rain)
Natural Environment	• Air, plant and animal • Components of soil • Housing and nature • Water cycle • Deforestation, decreasing dry land, increasing residential area in lowland and marshy areas, and land filling work on marshy place • Effect of chemical pollution to fish in Japan • Air and water pollution caused by waste • Air pollution caused by CO, CO_2, NO_2 • Acid rain, greenhouse effect • Causes and mechanism of ozone layer depletion • Effects of ozone layer depletion (plants, temperature increase) • Causes and mechanism of greenhouse effect (temperature, water cycle, polar ice) • Causes and mechanism of pollution caused by industrial gas (disease, air pollution, water pollution, acid rain)
Livelihood	• Mechanism of common colds, mumps, measles, polio, rabies, AIDS

shows the direct topics in the environment subject of the 9th and 10th grades according to the textbooks used (Curriculum Development Center, 2005c, 2005d). Mitigation measures, causes, the natural environment, and livelihood are the main direct topics, same as in science. However, this subject focuses on human activities. Environmental degradation is mentioned as the critical cause of disasters. To reduce damage, livelihood and environmental conservation are emphasized. The strong emphasis on

Table 4. Topics in the Environment Subject Directly Related to Disaster Management.

Mitigation	• Forest and controlling landslides • Mitigation measures (prevention of soil erosion, floods and landslides) • Safety measures at home and school • Mitigation of landslide and soil erosion • Planting tree, cleaning soil
Preparedness	
Response	• Importance of first aid
Recovery/Reconstruction	
Cause	• Lack of rain • Increase in atmospheric temperature • Desertification because of the excessive destruction of forests • Floods, landslides, droughts, desertification, excess rain, acid rain, storms and extension of sea-shores caused environmental degradation • Human activity, flood, landslide and soil erosion • Soil erosion, landslide, flood, deforestation, human activities and rain • Population, floods and landslide
Natural Environment	• Lack of rain • Increase in atmospheric temperature • Desertification because of the excessive destruction of forests • Floods, landslides, droughts, desertification, excess rain, acid rain, storms and extension of sea-shores caused environmental degradation • Forest and fresh air, forest and controlling landslides • Human activity, flood, landslide and soil erosion • Water, drought and environment • Deforestation, environment and landslide • Conservation of natural resources • Soil erosion, landslide, flood, deforestation, human activities and rain • Planting tree, cleaning soil • Drought, flood, landslide, soil erosion, desertification and development
Livelihood	• Human activity, flood, landslide and soil erosion • Human activity, soil, land and trees • Soil erosion, landslide, flood, deforestation, human activities and rain • Population, floods and landslide • Drought, flood, landslide, soil erosion, desertification and development

human activities is different compared to that in science. Moreover, problems (effects) are not shown in the textbooks.

Table 5 shows the indirect topics in the environment subject of the 9th and 10th grades. The number of indirect topics is more than that of direct topics. Nepal has numerous problems to solve. Such problems have the possibility to

Table 5. Topics in the Environment Subject Indirectly Related to Disaster Management.

Mitigation	• Mitigation measures (waste management, control water quality) • Mitigation measures for noise pollution (plant trees along the street) • Mitigation measures for waste water • Mitigation measures for waste management • Environment protection law • Role of government in conservation activities of forest • Role of housing in security • Measures for environment and public health (conservation of forests, pollution control, waste management, reduction in the use of chemicals, restoration) • Measures for ecosystem management
Preparedness	• Environment protection law • Role of government in conservation activities for forests
Response	• Disease • Lack of health services and facilities • Health services and facilities condition • Role of government and non-government agency in conservation • AIDS • Effects of air pollution (disease, body weak, damage of historical heritage, temperature increase) • Effects of land pollution (bad smell, health, damage of land condition, damage of beauty of environment) • Effects of waste water • Measures for population control
Recovery/Reconstruction	
Cause	• Human activity, soil, land, and tree • Land pollution • Causes of waste water production and disposal
Natural Environment	• Migration and environmental degradation • Environment and population • Climate and death • Climate and distribution of population • Supply of drinking water • Soil and water pollution caused by waste disposal • Effects of land pollution (bad smell, health, damage of land condition, damage of beauty of environment) • Development, land, water and forest • Air pollution and noise pollution caused by development activities • Population and deforestation • Industrial pollution and waste • Measure for environment and public health (conservation of forests, pollution control,

Table 5. *(Continued)*

	• Water pollution and human activity • Human activity, soil, land and trees • Air, water, land, minerals, forests, wildlife, climate as natural resources • Interrelationship between natural resources and human beings • Air pollution, water pollution and industry waste management, reduction in the use of chemicals, restoration) • Climate of Nepal • Geography of Nepal • Air on Earth
Livelihood	• Human activity and air pollution • Soil and water pollution caused by waste disposal • Water pollution and human activity • Interrelationship between natural resources and human beings • Air pollution, water pollution and industry • Air pollution and noise pollution caused by development activities • Population and deforestation • Industrial pollution and waste • Measures for environment and public health (conservation of forests, pollution control, waste management, reduction in the use of chemicals, restoration) • Sustainable development

cause disasters in some cases and disasters make Nepal's numerous problems more serious in some cases. More topics on the natural environment and livelihood are shown. Different from the direct topics, the effects (which belong to problems related to the response stage) are mentioned. Because this subject focuses on environment and human activities or livelihood, additional information or knowledge that can connect such topics and disaster management is necessary to include in disaster education. As an additional point differentiation of environment from science, the following topics are also mentioned in Table 5:

- Role of local bodies
- Role of health services and facilities condition
- Role of government and nongovernment agency for conservation

- Role of NGOs, clubs, and associations for community health
- Role of NGOs for measures of conservation/preservation of cultural heritage

The roles of organizations mentioned above are important in disaster education. But the linkage between these topics and disaster management is not mentioned in the textbook. Table 6 shows the direct topics in social studies in the 9th and 10th grades according to the textbooks used (Curriculum Development Center, 2005e, 2005f). The direct topics are less than in the other subjects. In social studies, the role of specific organizations in the response stage is described although only the role of the army and private companies are explained. Here, water-related disasters and soil erosion are mentioned. This is similar to what can be found in the environment subject. This subject is social studies, but the direct topics are similar to that of the environment subject because the topics are related to climate and human activities.

Table 7 shows the indirect topics in social studies in the 9th and 10th grades. The topics are similar to the topics in the environment subject

Table 6. Topics in the Social Studies Subject Directly Related to Disaster Management.

Mitigation	• Mitigation measures for hurricanes • Mitigation measures for soil erosion (tree planting)
Preparedness	
Response	• Response to hurricanes • Army provide emergency service during floods, landslides, earthquakes, air crash, mountaineering accidents • Role of private companies during drought
Recovery/Reconstruction	• Recovery from hurricanes
Cause	• Rain, flood, and drought • Links among economic activities, soil erosion, population growth, acid rain, partial rain, heavy rain, floods and landslides • Agriculture and soil erosion
Natural Environment	• Links among economic activities, soil erosion, population growth, acid rain, partial rain, heavy rain, floods, and landslides
Livelihood	• Links among economic activities, soil erosion, population growth, acid rain, partial rain, heavy rain, floods, and landslides • Agriculture and soil erosion

Table 7. Topics in the Social Studies Subject Indirectly Related to Disaster Management.

Mitigation	• Afforestation with the local village community • Mitigation measures for air pollution	
Preparedness		
Response	• Effects of polluted air (effects to people) • Causes of noise pollution and effects to people • Causes of water pollution and effects to people	
Recovery/Reconstruction		
Cause	• Causes of air pollution • Causes of noise pollution and effects to people • Causes of water pollution and effects to people	
Natural Environment	• Development and drinking water • Development, water, health, food, population • Mountain, river, and water • Environmental pollution • Climate • Deluge and ecological balance • Types of precipitation • Cyclone • Forest • Climate of Europe • Causes of air pollution • Effects of polluted air (effects to water, greenhouse effect) • Causes of water pollution and effects to people	• Housing, disease and drinking water • Links between housing and environment balance • Rain and agriculture • Rainfall • Factors affecting climate • Climate and vegetation • Climate and vegetation of Nepal • Geography of North America • Geography of South and Central America • Geography of Africa • River pollution • Environment and weather study
Livelihood	• Development and drinking water • Development, water, health, food, population • Drinking water distribution system • Electricity and communication as infrastructure • People's participation • Map around house • Map of school • Agriculture • Important measures that can develop agriculture • Industry in Nepal	• Measures to improve industry • Economy • Housing, disease, and drinking water • Link with housing and environment balance • Rain and agriculture • Housing • Human resource and development • Infrastructure • Community health • System of legislature • Map around house • Vehicle pollution

although social studies includes economic or industry aspects. In addition, because social studies also includes a geography chapter, many topics on the natural environment are involved. The roles or works of many organizations are written in the textbooks, but the linkage with disaster management is not shown. In addition, social studies pays attention to vulnerable people (with physical disability like blindness), but the linkage is also not shown.

Linkage between Disaster Education and School Curriculum

As for the disaster-related topics included in the textbooks, science includes the mechanism of disasters, causes, effects, and mitigation measures. Environment and social studies have a similar tendency with respect to topics, and these subjects cover links between the natural environment and human activities. Some mitigation measures are shown in the textbooks, but they are not emphasized. These two subjects focus on mechanism and causes of disasters, especially floods and landslides. Mitigation measures are described in the textbooks in the three subjects, but they are mainly at the individual and community levels. The role of government or other institutions cannot be identified. In addition, the description of past disasters is not shown. The main topics are mitigation measures at the individual level, causes of disasters, social environment including human activities, and natural environment. At the response stage, some topics are shown in the textbooks, but they are limited.

By learning these three subjects, the mechanism, causes, and effects of disasters and the links between the natural and social environments can be provided as part of disaster education. However, school disaster education in Nepal does not provide information or knowledge on past disasters or on the roles of stakeholders. But when disasters occur, the impacts are seen in various aspects: on the natural environment, people, government, economy, livelihood, etc. In the local community, people have their own livelihood concerns or problems to be solved. But school disaster education in Nepal does not provide knowledge on such things even if they are the most crucial parts of school disaster education.

However, there are many topics that have the possibility to become disaster-related topics in textbooks (the indirect topics discussed earlier). If the indirect topics are provided as disaster education, it is necessary to indicate the links between these topics and hazards or disaster management. For example, recently, climate change adaptation is one of the big challenges in disaster management. But if teachers teach about climate, it becomes environment education, which is part of the environment subject in the case of Nepal. But if the phenomenon of climate change is shown as the cause of disasters, it becomes disaster education. As another example from social studies in Table 7, there is the topic of housing. If students think about housing from aspects of daily life and emergency situation, the housing topic can be part of disaster education. Thus, the school can increase the kinds of education topics by using indirect topics, and such disaster education is more related to society and daily life.

Science textbooks mention disaster-related topics clearly, but the range of disaster education is limited compared to in environment and social studies. Environment and social studies have the possibility to increase the opportunities of school disaster education. These two subjects are significant for future school disaster education. In this section, three subjects are examined as disaster-related subjects. But any number of subjects can actually be utilized if disaster-related topics are included (for example, disaster statistics are sometimes discussed in mathematics) or if the teachers can create such topics. The main problems of the curricular education in Nepal in terms of disaster education are the following:

- The roles of the government or other institutions in any stage of the disaster management cycle are not shown comprehensively. It is difficult for students to know the specific problems encountered by the government or by other agencies.
- The lessons, problems, experiences, and measures in the past disasters are not shown.
- The activities in the recovery stage are not included.
- The disaster management cycle is not explained.

Hopefully these shortcomings can be addressed in the future textbooks of the students in Nepal.

INTEGRATING DISASTER EDUCATION INTO SCIENCE CLUB ACTIVITIES IN THE PHILIPPINES

A survey conducted by a public opinion polling body found that 10 percent of the Filipino youth were connected to a youth organization (ESCAP, 2000). Six percent were members of arts, music, or educational organizations; three percent were part of a charitable organization; and one percent of the youth were affiliated to a political party, labor union, or professional association.

Some of the most popular youth organizations, especially in the primary and secondary schools, are the science clubs. In the Philippines, there are two major networks of science clubs. The first one is the Philippine Society of Youth Science Clubs (PSYSC), a 40-year-old NGO registered with the Securities and Exchange Commission, while the second one is the Youth for Environment in Schools Organizations (YES-O), which is coordinated by a government agency attached to the Department of Education (DepEd).

Philippine Society of Youth Science Clubs (PSYSC)

PSYSC was established on March 5, 1971, by 26 science fair winners, making it one of the oldest youth-run science organizations in Asia (PSYSC, no date). It is a not-for-profit NGO whose main objective is to promote the public understanding of science, technology, and the environment (ST&E). PSYSC's annual membership of more than 100,000 includes elementary school pupils, high school students, and science and math teachers. PSYSC is managed entirely by volunteer university students. The age of the volunteers ranges from 16 to 23. Majority of the volunteers come from the University of the Philippines, the country's premier learning institution. The goals of PSYSC are to initiate programs, projects, and activities that shall complement and supplement formal science education; to popularize science clubbing as an effective means of promoting scientific consciousness and scientific culture among the youth; to initiate, maintain, and monitor a network of youth science organizations; and to call for the involvement of public and private sectors in ST&E programs, projects, and advocacy. To achieve its goals, PSYSC conducts the following major activities:

The National Youth Science, Technology, and the Environment Summer Camp

Every year the camp draws around 800 participants from all over the country. The participants are high school students, aged 12–16, and their science teachers. The camp, a four-day activity, is usually held at the end of March or in early April. The venue changes every year, alternating between the three major island groups of the Philippine archipelago (Luzon, Visayas, and Mindanao). Camp 2010, the most recent camp, was organized by PSYSC in cooperation with Kyoto University, the Asian University Network of Environment and Disaster Risk Management (AUEDM), the United Nations International Strategy for Disaster Reduction (UNISDR)

Bangkok Office, the United Nations Educational, Scientific, and Cultural Organization (UNESCO) Jakarta Office, and the Philippine Institute of Volcanology and Seismology (PHIVOLCS). The camp theme was "Alert Level '10: The RESCUE (*R*esilient, *E*mpowered *S*cience *C*lubs even *U*nder *E*mergency)." Camp 2010 was about rethinking the way young people prepare for hazards and respond to disasters. Lectures included "Science of Disasters and Detection of Hazards" (explanation on the different kinds of geological and hydro-meteorological hazards; recent major disasters in the Philippines; advances in Doppler radar and seismograph technology and other ways of knowing when a disaster is impending); "Community in Action: From Tragedy to Strategy" (impact of disasters on national development; community-based disaster risk management); and "Disaster Preparedness 101" (4 Rs of Disaster Management: Reduction, Readiness, Response, Recovery; thinking globally, acting locally: what science clubs can do to help in building the disaster resilience of their community). Camp 2010 provided a venue for high school students and teachers to meet real scientists and experts in the field of Disaster Risk Management, like Dr. Renato Solidum, the Director of PHIVOLCS, and Mr. Nathaniel "*Ka Tani*" Cruz, the former Chief of the Weather Division of the Philippine Atmospheric, Geophysical, and Astronomical Services Administration (PAGASA). Advocating fun experiential learning, PSYSC created a venue that allowed scientists to interact directly with students, enabling them to make valuable contributions like modeling scientific inquiry, sharing their passion for science, connecting science and technology to the "real" world, and augmenting the science background of the teachers and students.

The direct beneficiaries of Camp 2010 were the students and teachers who attended the event. The indirect beneficiaries were the households, science clubs, schools, and communities that they belong to and those who stand to benefit from the knowledge and experience the students and teachers gained from their participation in the camp. For example, as a pre-camp workshop, all participating science clubs were required to perform community mapping, identifying hazardous areas, safe areas, and evacuation routes (see Fig. 3). The exercise aimed to bring the attention of the students and teachers on their local surroundings, which are often taken for granted. Practical activities like this were taught in the camp and can be replicated by the students and teachers when they return to their respective schools.

The National Children's Science Interactive Workshop (CSIW)

CSIW is the camp designed for PSYSC's elementary school affiliates aged 9–11. CSIW was conceptualized to address the need to encourage

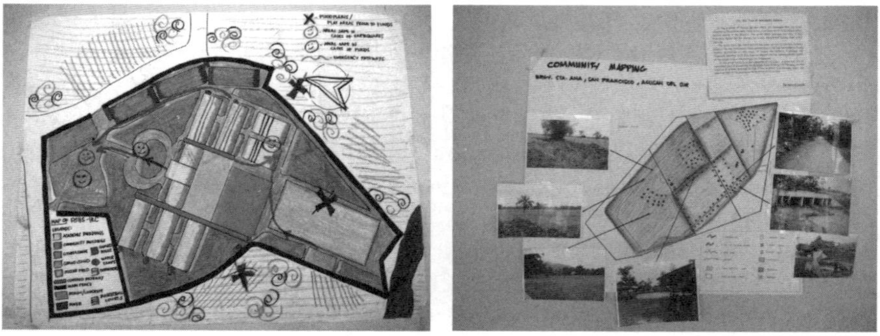

Fig. 3. The Community Maps of the Philippine Science High School – Ilocos Region Campus (left) and of Barangay Sta. Ana in San Francisco, Agusan del Sur (photos taken by author).

elementary pupils' interest in science. CSIW 2010's topic was biodiversity, coinciding with the celebration of 2010 as the International Year of Biodiversity. In the group talent competition the students showcased the importance of biodiversity through song and dance. Most of the performances linked destruction of the environment like deforestation to the loss of biodiversity and to the recent destructive disasters experienced by some provinces like landslides and floods (see Fig. 4).

The Collaborative Undertaking to Revive the Environment (CURE)

PSYSC, with the help of the Department of Environment and Natural Resources (DENR), various Rotary Club chapters, corporate sponsors, and local government units, launched CURE. Elementary and high school science club members engage in cleanup activities and tree planting. There are also awareness seminars by resource persons from DENR. Here the speakers discuss Principles of Ecology, Protected Area Management, Watershed Management, etc. In the school year 2005–2006, five established regional councils of PSYSC were able to implement CURE activities in their respective regions. The CURE that year was aptly given the theme "Bridging Initiative Geared Strongly Towards Earth's Protection" (B.I.G. S.T.E.P.). These three major activities are easily replicable by member clubs throughout the country. PSYSC believes that the youth should not be left out in disaster preparedness activities. Young people are a crucial segment of society. Given proper guidance, the youth can contribute actively in community safety. It is important to strive toward the involvement of young people as active citizens.

Fig. 4. Elementary School Students Participating in CSIW 2010 Dramatized through Song and Dance the Bad Impacts of Rampant Deforestation on Biodiversity and on the Safety of People Living at the Foot of Mountains (photo taken by author).

Youth for Environment in School Organizations (YES-O)

In 2003 the Center for Students and Co-Curricular Affairs (CSCA) of the Department of Education (DepEd), in cooperation with the Department of Environment and Natural Resources (DENR) Youth Desk, pushed for the establishment of Youth for Environment in Schools Organization (YES-O) as "a school-based co-curricular organization which will serve as a significant venue for students' actions and movements toward safeguarding, protecting, and conserving the environment for future generations" (DepEd, 2003). Coincidentally, the 2010 Youth for Environment Summer (YES) Camp organized by CSCA-DepEd and DENR Youth Desk, in collaboration with the NGO *Kabataang Sama-samang Maglilingkod* (Youth Unified in Service), Inc. (KASAMA), also had a disaster risk reduction theme. The camp was held in Baguio City and the theme was "*Sagip Kalikasan, Sagip Buhay*" (Disaster Management through Environmental Protection). One of the objectives of the YES Camp was to "inculcate among students the critical role of environmental protection in preventing and coping with the effects of natural disasters such as typhoons, floods, earthquakes, and droughts" (DepEd, 2010). Activities during the YES Camp included interactive plenary sessions, panel discussions, action planning, outdoor activities, contests (quiz, debate, poster-making, documentary, etc.), and exhibits of success stories on environmental action and initiative.

Two Important Months for Science Clubs and Disaster Education

Although science club activities related to disaster risk reduction, climate change adaptation, or disaster education can happen anytime during the school year, there are two special months when more conspicuous and elaborate activities are organized, sometimes as directed by the DepEd head office through formal department orders and memoranda.

The National Disaster Consciousness Month (July)
In August 1999, Executive Order No. 137 was issued by former Philippine President Joseph Estrada to declare the month of July of every year as the National Disaster Consciousness Month (NDCM) in order to heighten public awareness on the importance of disaster prevention, mitigation, and preparedness. Annually, the Department of Education enjoins all its offices, private and public schools, and attached agencies to undertake relevant activities to mark the significance of the month-long event. In 2010, the NDCM theme was "*Safe Ka Ba? Programa Laban sa Kalamidad Tungo sa Pag-unlad*" (Are You Safe? Program Against Disasters Toward Progress). The DepEd Secretary ordered schools and all DepEd offices to hang streamers reflecting the theme; share lessons on DRR during flag-raising ceremony; engage in the discussion of disaster prevention, mitigation, and preparedness in class; advocate DRR as a way of life; hold a symposium on DRR; and conduct essay writing and slogan contests on disaster preparedness (Cabasal, 2010). Science clubs participated in the celebration by conducting their own activities, like organizing tree planting, cleanup drives, film-showing about climate change and global warming, blogging about DRR, etc.

The National Science Club Month (September)
In 1979, President Ferdinand Marcos signed Presidential Decree No. 1895 declaring September of every year as the National Science Club Month (NSCM). PSYSC, the Science Club Advisers Association of the Philippines (SCAAP), CSCA-DepEd, and the Science Education Institute of the Department of Science and Technology (DOST-SEI) lead the annual celebration. PSYSC has long been taking a very active part in the success of

the annual NSCM celebration. The organization had launched the following contests in recent years: the PSYSC Science Olympiad (where they included questions about different geological and hydro-meteorological natural hazards under Earth Science) and the mathematics, science, *at Kapaligiran* (and the environment) (Mathsciaka) Contests (where they included a workshop on water filtration using locally available materials, which can be useful during disasters, if supply of clean water is suddenly cut off). These two months of each year, July and September, offer opportunities for further integration of disaster education into science club activities. The annual NDCM and NSCM celebrations make sure that disaster education in schools is regular and, in a way, sustainable as it is backed by two presidential orders that are fortunately being observed religiously.

The Potential of Science Clubs as Vehicles for Disaster Education

The involvement of science club members and advisers in DRR is important because they know very well the conditions of their school and community and have knowledge about locally available resources for a potentially sustainable community-based disaster risk reduction (CBDRR).

As shown in Fig. 5, the engagement of science clubs can be increased by the participation in multisectoral DRR processes of external actors like science club coordinators (PSYSC and CSCA-DepEd), which implement annual nationwide activities related to the NSCM and NDCM celebration

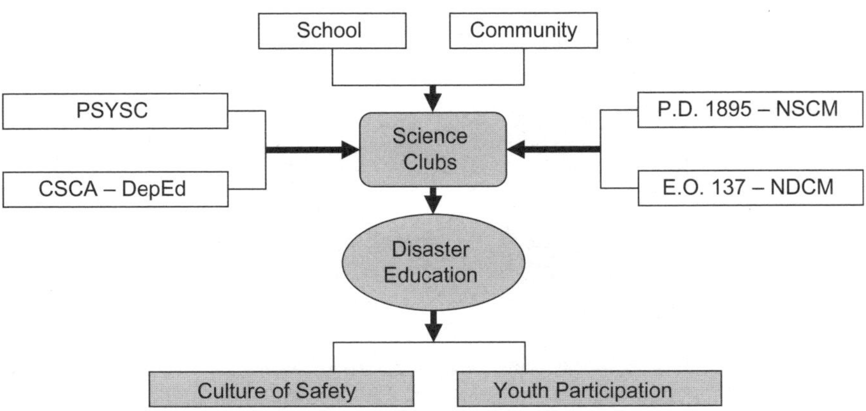

Fig. 5. Science Clubs as Vehicle for Disaster Education.

mandated by law. Given their wide distribution across the country and their ability to integrate DRR in their activities, science clubs are appropriate vehicles for disaster education. Active science clubs can therefore contribute enormously to building a culture of safety and enhance youth participation in DRR. Young people need organizations, like science clubs, that provide positive, supportive environments. The youth need to have opportunities to contribute to society. Young people have great promise in addressing environmental concerns like DRR because of their energy and passion. To enhance youth participation in DRR, they must be provided with Disaster Education, both in school and out of school. Science clubs are an excellent example of how to supplement formal disaster education in the classroom.

CONCLUSION

This chapter showed three cases of school disaster education:

- Innovative concept and programs of school disaster education
- Possibility of integration of disaster education and school curriculum
- Integration of school disaster education between inside and outside

Fig. 6 shows the concept of school disaster education proposed in this chapter. In the figure, curricular education means education through the use

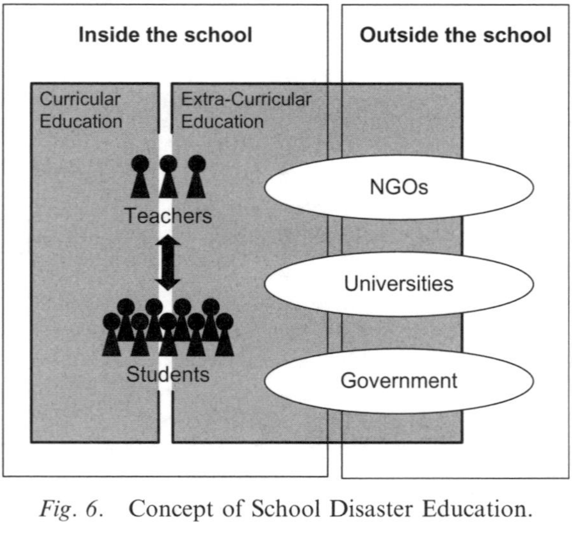

Fig. 6. Concept of School Disaster Education.

of regular subjects like chemistry, mathematics, social studies, and others. Therefore, curricular education is generally conducted inside the school. In the case of extra-curricular education, it is possible to conduct it inside and/or outside the school. Utilizing more opportunities in school education contributes to enhancing the variety and effectiveness of school disaster education. In case of education inside the school in extra-curricular education, the school can invite related organizations/resource persons to impart knowledge to students as well as to facilitate the students coming up with their own programs. In the case of education outside the school, the students can learn about and participate in the activities organized or supported by other organizations. To enrich school disaster education, it is important to explore how the school can provide disaster education in the current school curriculum and what kinds of disaster education programs can be provided both inside and outside the school in collaboration with NGOs, universities, the government, communities, and other related organizations/individuals. The school is not only a place for providing disaster education but also a key institution that can arrange disaster education with the help of external organizations.

ACKNOWLEDGMENT

This chapter compiles the results of different researches conducted by the first author. Mr. Seiji Suwa of Maiko High School provided information about the Environment and Disaster Mitigation Course. Some students of the course were kind enough to be interviewed about their experience in this new kind of education. NSET-Nepal cooperated in the survey to examine the linkage between school curriculum and disaster education. They are all gratefully acknowledged. We would also like to extend our gratitude to the Ministry of Education, Culture, Sports, Science, and Technology (MEXT) of Japan for second author's scholarship at Kyoto University and to the Environmental Management Leaders (EML) Program for funding the internship programs.

REFERENCES

Arya, S. A. (1993). Training and drills for the general public in emergency response to a major earthquake. *Training and Education for Improving Earthquake Disaster Management in Developing Counties, UNCRD Meeting Report Series*, 57, 103–114.

Cabasal, M. (2010). DepEd Celebrates National Disaster Consciousness Month. Department of Education. Pasig, Philippines. Available at http://www.deped.gov.ph/posts.asp?dp=64

Curriculum Development Centre. (2005a). Science, Grade 9. His Majesty's Government Ministry of Education and Sports, Curriculum Development Centre, Nepal.

Curriculum Development Centre. (2005b). Science, Grade 10. His Majesty's Government Ministry of Education and Sports, Curriculum Development Centre, Nepal.

Curriculum Development Centre. (2005c). Health, Population & Environment, Grade 9. His Majesty's Government Ministry of Education and Sports, Curriculum Development Centre, Nepal.

Curriculum Development Centre. (2005d). Health, Population & Environment, Grade 10. His Majesty's Government Ministry of Education and Sports, Curriculum Development Centre, Nepal.

Curriculum Development Centre. (2005e). Social Studies, Grade 9. His Majesty's Government Ministry of Education and Sports, Curriculum Development Centre, Nepal.

Curriculum Development Centre. (2005f). Social Studies, Grade 10. His Majesty's Government Ministry of Education and Sports, Curriculum Development Centre, Nepal.

Department of Education (DepEd). (2003). Establishment of the Youth for Environment in Schools (YES) Organization. DepEd Order No. 72. Available at http://www.deped.gov.ph/cpanel/uploads/issuanceImg/DO%2072_09-01-03_00001pdf

Department of Education (DepEd). (2010). 8th Youth for Environment Summer (YES) Camp. DepEd Memorandum No. 62. Available at http://www.deped.gov.ph/cpanel/uploads/issuanceImg/DM%2062%20S%202010.pdf

Dixit, M. A. (2004). Challenges of implementing school earthquake safety programs in Nepal. Fourth Annual IIASA-DPRI Meeting: Integrated Disaster Risk Management: Challenges of Implementation. Available at http://www.iiasa.ac.at/Research/RMS/dpri2004/Abstracts/abs_dixit.pdf

Economic and Social Commission for Asia and the Pacific (ESCAP). (2000). *Youth in the Philippines: A review of the youth situation and national policies and programmes.* York: UN.

Frew, L. S. (2002). Public awareness and social marketing. ADPC Regional Workshop on Best Practices in Disaster Management, Bangkok, 381–393.

Izadkhan, Y. (2004). Risk education: Road to community protection. Fourth Annual IIASA-DPRI Meeting: Integrated Disaster Risk Management: Challenges of Implementation. Available at http://www.iiasa.ac.at/Research/RMS/dpri2004/Abstracts/abs_yasamin.pdf

Kuroiwa, J. A. (1993). Peru's national education program for disaster prevention and mitigation (PNEPDPM). Training and Education for Improving Earthquake Disaster Management in Developing Countries, UNCRD Meeting Report Series, 57, 95–102.

Maiko High School. (2002). Educational goals. Available at http://www.hyogo-c.ed.jp/~maiko-hs/e/EDM_course/goals.htm. Retrieved on December 22, 2010.

Maiko High School. (2010). Volunteer activities. Available at http://www.hyogo-c.ed.jp/~maiko-hs/. Retrieved on December 6, 2010. [in Japanese].

Philippine Society of Youth Science Clubs. (No date) PSYSC history. Available at http://www.psysc.org/profile.html

Radu, C. (1993). Necessity of training and education in earthquake-prone country. Training and Education for Improving Earthquake Disaster Management in Developing Countries, UNCRD Meeting Report Series, 57, 15–33.

Shaw, R., & Kobayashi, M. (2001). The role of schools in creating earthquake-safer environment. Paper presented at the OECD Workshop Disaster Management and Educational Facilities, Thessaloniki, Greece, 7–9 November 2001.

Shaw, R., Shiwaku, K., Kobayashi, H., & Kobayashi, M. (2004). Linking experience, education, perception and earthquake preparedness. *Disaster Prevention and Management, 13*(1), 39–49.

Shiwaku, K. (2004). *Study on disaster education at high school: Role of education of the environment and disaster mitigation course at the Maiko High School.* Master thesis, Kyoto University. [in Japanese].

Shiwaku, K. (2007). *Towards innovation in school disaster education: Case research in Kathmandu, Nepal.* Ph.D. thesis, Kyoto University.

Shiwaku, K., & Shaw, R. (2008). Proactive co-learning: A new paradigm in disaster education. *Disaster Prevention and Management, 17*(2), 183–198.

Shiwaku, K., Shaw, R., Kandel, R. C., Surya, S. N., & Dixit, A. M. (2006). Promotion of disaster education in Nepal: The role of teachers as change agents. *International Journal of Mass Emergency and Disasters, 24*(3), 403–420.

Strike, A. K. (2000). School as communities: Four metaphors, three models, and a dilemma or two. *Journal of Philosophy of Education, 34*(4), 617–642.

UN/ISDR. (2007a). 2006–2007 World Disaster Reduction Campaign. Available at http://www.unisdr.org/eng/public_aware/world_camp/2006-2007/wdrc-2006-2007.htm

UN/ISDR. (2007b). 2006–2007 World Disaster Reduction Campaign: Disaster risk reduction begins at schools. Available at http://www.unisdr.org/eng/public_aware/world_camp/2006-2007/pdf/WDRC-2006-2007-English-fullversion.pdf

Wisner, B., Kelman, I., Monk, T., Bothara, L. J., Alexander, D., Dixit, M. A., Benouar, D., Cardona, D. O., Kandel, C. R., & Petal, M. (2004). School seismic safety: Falling between the cracks? Fourth Annual IIASA-DPRI Meeting: Integrated Disaster Risk Management: Challenges of Implementation. Available at http://www.iiasa.ac.at/Research/RMS/dpri2004/Papers/Ben_SchoolSafety.pdf

CHAPTER 4

ROLES OF FAMILY AND COMMUNITY IN DISASTER EDUCATION

Yukiko Takeuchi, Farah Mulyasari and Rajib Shaw

INTRODUCTION

Generally, family and community have a great deal of local experience and local knowledge of disaster. Disaster education for family and community is aimed at recognizing the characteristics of a disaster and the existing social situation for the purpose of acquiring general knowledge of disasters, usually at school. Community and family structures and roles differ according to character and location such as urban, rural, coastal, near rivers, and near mountains, among others. In recent times, people's participation in the community has been affected by social changes. Earlier, historical local disaster prevention methods were passed on to other family/community members through daily activity. Recently, however, the characteristics of disasters have changed such that people now need to prepare for disasters of which they have no experience and about which they have difficulty obtaining information. It is thus necessary for communities and families to know different scenarios of disaster. "Community-Based Disaster Risk Management" is difficult to establish without linking community and household. For instance, many types of associations can be found in the community, but some do not play a direct role in disaster prevention and management. However,

Disaster Education
Community, Environment and Disaster Risk Management, Volume 7, 77–94

ISSN: 2040-7262/doi:10.1108/S2040-7262(2011)0000007010

these associations have strong human relationships and much local knowledge. As an example, family members traditionally take care of children and old and handicapped people. Presently, lifestyles and social systems have changed such as long-distance commuting, both husband and wife working, weak family relationships, fewer children, aging, and unstable economies, among others. It is therefore necessary to carry out disaster prevention education aimed at building local capacity for disaster prevention, after determining the situation in the community and family and the roles people in the community play.

Disaster education in family and community could be "external" and/or "internal." "Internal" disaster education within families and communities refers to the sharing of past experiences (e.g., religion, traditional and indigenous knowledge, local wisdom, and customs or living habits) among the members in the family and community. In particular, the "religion" of a family influences the behavior of individuals as well as their perception of risks. Traditional and indigenous knowledge (i.e., technology, know-how skills, practices and beliefs of the local community), accumulated over generations based on local experience and historic reality, has specific local context and may be embedded within the individuals, families, and members in the community, which in turn, influences their perception of risks and response to hazards. Another element that characterizes the disaster education in the family and community is the nature of the community itself and where the families are living. For example, the community is composed of various groups of people; young, elderly, female-dominant or male-dominant, which in turn has an effect on shaping the disaster education and learning process within family and community.

The above-mentioned elements and key issues illustrate the involvement of family and community in disaster education. The next section explains in detail types of families as well as the linkages of disaster education to families and communities. Finally, the chapter provides specific educational aspects suitable for different types of families and communities.

LINKAGE OF DISASTER EDUCATION TO FAMILIES AND COMMUNITIES

Petal (2009) suggested that the mission of education about disaster is to convey and understand the natural and environmental conditions as well as the human action and inaction that lead to disaster, to stimulate changes in individual and group behavior, and to motivate advocacy and raise expectation of social policy to reduce these threats with the involvement of all stakeholders. It is important to note that disaster education should not be an event (like an evacuation drill), but should be a continuous process. Furthermore, disaster education should not be limited within the school, where education is provided, but should break the school boundary and be linked to the community and family. The importance of linking school education with family and community education has been gradually recognized and currently is practiced in several countries, which will be explained in detail in a later section. Family education is related to education obtained through the interactions with and among the family members, through daily activities and conversations. Community education may be obtained through participating in voluntary and community activities, drills, seminars, and other public activities.

Individuals, families, and communities vary and their characteristics differ between urban (Fig. 1) and rural areas (Fig. 2). Fig. 1(a) illustrates an elderly person living in an urban setting. The elderly person lives alone, disconnected from the community, likely lacks the ability to self-help during emergencies, and is highly vulnerable during an emergency. The elderly person may be equipped with knowledge acquired from previous experiences, however, due to poor social networks and lack of interaction with the community, the knowledge is not transferred. Fig. 1(b) illustrates university students and working people who live alone in single-room apartments. This group of individuals would have the ability to self-help and may be able to provide assistance to others during emergencies. However, due to hectic schedules, they may not be able to participate in community activities or interact with neighbors, thus there is poor social network and they may not feel the need to help neighbors during emergencies. Fig. 1(c) illustrates a typical household, consisting of parents and children. There

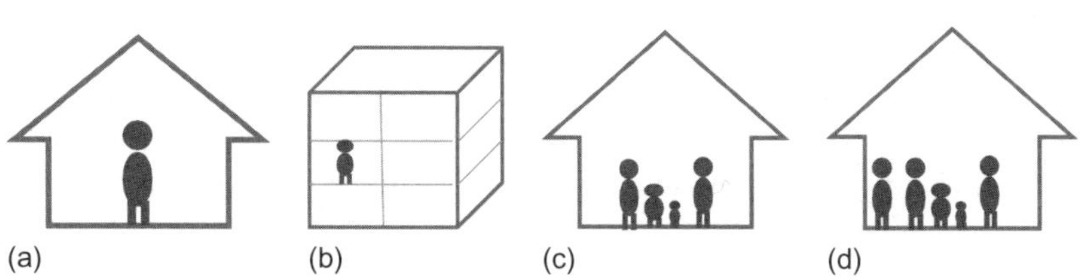

Fig. 1. Various Types of Individuals and Members within a Family and Community.

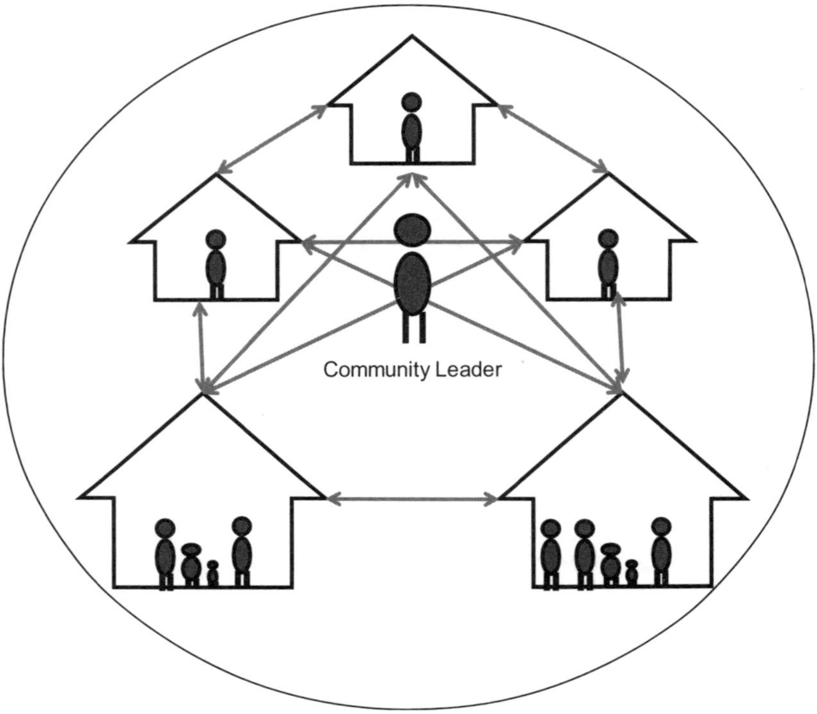

Fig. 2. Individuals and Characters within a Family and Community in Rural Area.

may be community communication and participation through the children's school activities. Furthermore, the parents would have the ability to self-help and provide assistance to others when in need. In families where the parents have experienced past disasters, they would then be able to share these experiences with the children, which could serve as a form of disaster education within the family. Fig. 1(d) illustrates a household consisting of not only the parents and children but also the grandparents. Similar to the case in Fig. 1(c), there would be interactions with the community via school events and the parents would have the ability to self-help and provide assistance for others in emergencies. In terms of family education on disaster, in addition to parents, the grandparents would also be able to share their experiences and lessons with the younger generations.

Fig. 2 shows the families that make up the community in a rural setting. The community consists of a mixture of households (i.e., young and elderly

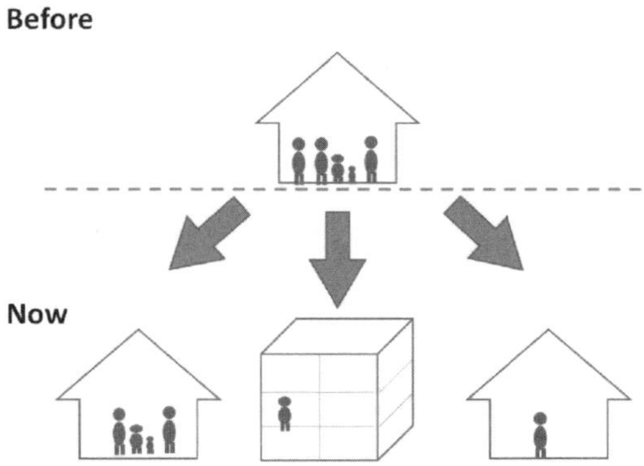

Fig. 3. Living Condition of Family in a Community in Japan Before and After the 1970s.

people, single-headed families, etc.). Key elements that influence disaster education would be the awareness of local environment (i.e., geology and topography), traditional and indigenous knowledge, social networking, and past experiences. In general, rural communities have a leader who is trusted by the local residents, who is the linking factor, and who plays the key role in disseminating information and mobilizing resources during emergencies.

Development also has an effect on the type of living conditions, which in turn, affects the type of disaster education available. Fig. 3 shows the household make-up before and after the 1970s. Before the 1970s, it was common for a household to consist of a mix of several generations. The senior members, who have experienced disaster events, had direct contact with the rest of the family and could transfer their knowledge to the younger generations. Unfortunately, with modernization, individuals tend to move out, live alone in apartments, and young couples start their own families. The disaster education learning process is disconnected.

Having explained the various forms of family, their respective disaster-education linkages and shortcomings, Fig. 4 illustrates the ideal scenario of disaster-education linkages between families and communities. There will be three types of disaster-education communication among the family members, namely: i) between the parents; ii) between parents and children; and iii) between the senior members and the younger generations. There will also be communication between the family and community, i.e., the senior members communicate with the networks in the community and the children interact with the others at school.

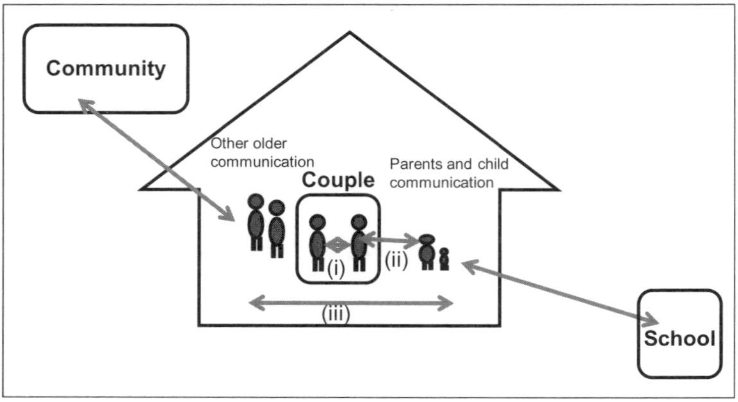

Fig. 4. Disaster Education Linkage in a Family to the Community.

IMPLEMENTATION OF DISASTER EDUCATION IN FAMILIES AND COMMUNITIES

Understanding the patterns and behavior of social organizations within society such as families and communities, it is essential to promote positive perception and action for disaster education. Societies might differ in culture, values, customs, and also in level of development, but the social units remain more or less the same. Similar disaster events may cause different consequences depending on the region and vulnerability (i.e., sex, age, ethnic group, caste, etc.). Due to differences in vulnerability, behavior and strategies toward disaster education and disaster risk reduction varies. Details will be discussed based on the case studies in later sections. The social units, which are affected by disasters, refer to the families and communities, whose role in disaster education has been discussed. It is necessary to consider how these units operate, separately and as a composite whole, the potential problems, and their contributions to disaster education.

School plays a key role in disseminating disaster education in the community. Shiwaku (2009) argued that the importance of school disaster education is increasing, because children are one of the most vulnerable groups within the society; they are the future generation; the school is the center of the education, and effects of education can be transferred to parents (families) and community. School also functions as a cultural and spiritual center for the community and a facility for education. Through these interactions, linkage between members of the families and community can be built. It can also provide an opportunity for disaster education to be transferred among the community.

Ensor (2010) pointed out that education is increasingly accepted as an integral part of humanitarian response in emergencies. Access to good quality education in disaster situations has proven to be critical to the protection of children as well as the well-being of their families and communities. It was further emphasized that instead of considering children and women as the victims and/or the vulnerable group, they should also be recognized for their positive roles in facilitating disaster education, preparedness, and response during emergencies. Women are important communicators to their families and can mobilize the participation of the whole community. Nimpuno (2007) suggested the following when considering ways of involving women in disaster preparedness and education, namely: i) raising survival awareness through organization of activities such as emergency health measures for survival and vulnerability reduction; and ii) enhancing women's ability to meet local needs for disaster reduction and utilization of women's skills for disaster preparedness. In many countries, women play crucial roles in areas such as health and sanitation as a part of disaster preparedness and risk reduction education. Takeuchi and Shaw (2009) emphasized that risk communication and education to women is an absolute requirement for preparedness. Furthermore, women/female participants in community meetings are strongly encouraged. Such involvement could be achieved by linking participatory education among the school children, such as neighborhood/town watches as well as Parent-Teacher Associations (PTAs).

In the case of the implementation of disaster education at the community level, the Community-Based Disaster Risk Management (CBDRM) aims at strengthening community capacity (i.e., individuals, families, and communities) so as to reduce vulnerability and disaster risks (Victoria, 2009). As emphasized by Victoria (2009), understanding people's coping mechanisms and recognizing and strengthening community capacities is important in the process of disaster education. Thus, the CBDRM approach is people-oriented with the community members being the main actors. Disaster education focuses on traditional knowledge, local resources and practices, skills, support, and structures for disaster preparedness and mitigation at the local context. In addition, it promotes the building of social networks within the community. Nimpuno (2007) suggested that social networks help the accomplishment of disaster tasks. For example, traditional social networks may be used as a communication tool for disseminating disaster-related information, and in turn, this serves as a local warning system. These social networks perform important socio-economic functions and if properly mobilized, they can act as a significant force in disaster education. In spite of the variety of networks, they are very much interlinked with mainstream society. It is essential to identify the network within the community to perform the appropriate functions so as to achieve maximum effectiveness and efficiency.

CASES OF DISASTER EDUCATION IN FAMILIES AND COMMUNITIES

This section describes examples of the disaster education implementation in families and communities in selected Asian countries, namely Japan, India, and Indonesia, well-known for its resilience toward disaster risks.

The Case of Japan

Japan experienced rising economic growth from 1955 to 1975. The urban area spread under the influence of economic growth with a corresponding population increase. Thus, residential areas expanded in the urban area. These locations were former fields and near mountains or rivers. The area's residents do not have a deep relationship with other residents and do not know local history and indigenous knowledge. For some, it is their first experience of living near mountains or rivers. Often, many do not understand the disaster situation. Hiroshima city is one such large city in Japan. One case city experienced serious rainfall disaster in 1999. Hiroshima city is a large city but, like other coastal cities in Japan, it lacks sufficient flat land. Between the 1970s and the mid 1990s, there has been a steady population growth and people from different places have settled in the city. As a result of this growth, the city has spread out over the mountain area. From past experiences and memory, older residents know where the risky parts are and when the disaster may happen, but new residents do not have that knowledge. On 29 th June 1999, due to an atmospheric depression, heavy rainfall (over 150 mm/3 hours) occurred in Hiroshima city. Thirty-two people died in this 1999 disaster. One victim, a 29-year-old woman who died inside her house, is illustrative of the situation. The place where she died is a high-risk debris flow area. In the 1970s, this area was a paddy field. After the 1970s, the land use changed to residential and a new transport system was developed in the 1990s. It grew into a large residential area and by the end of the 1990s, the pond was reclaimed and 10 houses were built on it. Additionally, it is located at the entrance of the valley and prone to debris flow. Many new residents moved to this area. One young leady died in this disaster. Her family started living in this area 3 years before the 1999 disaster. Her family comprised her husband and two children. On the day of the disaster, her husband went to work while she stayed at home with the children. This family did not have many contacts in this community and was not included in the community network. The community network was a social network consisting of members of the Local Voluntary Disaster Management Organization and of the residents' association, which exists in each community. On the day of the disaster, there was an official announcement at 10 am to inform residents through TV, radio, and the community network of the dangers of heavy rainfall, but this family was not part of the community network. Around 12 pm residents received further information about the changing color of river water, rumbling from the mountain, and a smell of burning, among others. For this woman, it was the first experience of living in a mountain area. She did not understand that the information was a prelude to debris flow and many flows eventually occurred between 1 pm and 3 pm. In this area, communities have Local Voluntary Disaster Management Organizations, and every year, they hold evacuation training programs. However, the training participants are male in the majority of cases. Her husband learned some risk information and other local information in this way but did not share it with other family members, and he left the community during daytime. She did not know about the evacuation shelter procedure in this community and stayed in the house. When the debris flow happened, she stayed on the first floor; her two children stayed on the second floor. One debris flow directly hit her house. The children were rescued from the second floor but the dead woman was pulled from the debris after 6 hours (Takeuchi & Shaw, 2009).

Japanese life style was changed by long commuting distances, both husband and wife working, weak family relationships, fewer children, and an aging society. Some areas have a strong community network and relationship. However, in many parts, it is difficult to make relationships with the community and spend time for disaster prevention. Disaster education for Japanese people needs to encompass self-help, collaboration with community, and collaboration with the public. There can be four types of relationship based on the local characteristics. Ideally, Type D of Fig. 5 is the best one, where there is an overlap of self-help, mutual help, and public help. However, in many cases, there is no overlap of self-help in individual and self-help in family (Type A). In some cases, there are overlaps of

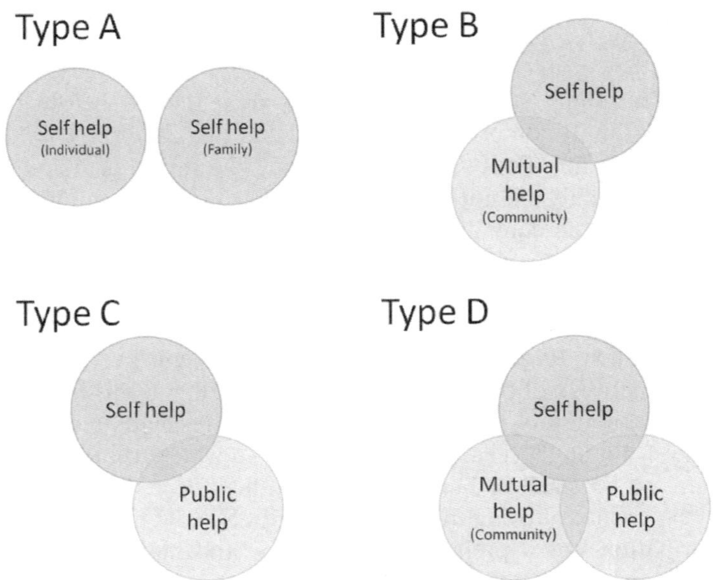

Fig. 5. Types and Characteristics of Self-help, Mutual Help, and Public Help.

self-help and mutual help (Type B), where the local communities and family-individual relations are strong. In other cases, there is no overlap of self-help and mutual help, but people depend on public help, and therefore Type C is important in those cases. Based on these local characteristics, different types of education are important and the role of family and community becomes different in each case. Fig. 6 shows the educational needs and types based on the local characteristics, which is related to Fig. 5. It shows that risk information needs to include the natural and social condition, local vulnerability, and information from the local government. Based on the local characteristics and social dynamics, preparedness types will be different for either preparation by oneself, family, community, public association, and/or NPO.

The Case of India

In 2004, many people from different countries were affected by the Indian Ocean tsunami. Especially in coastal areas, people sustained serious

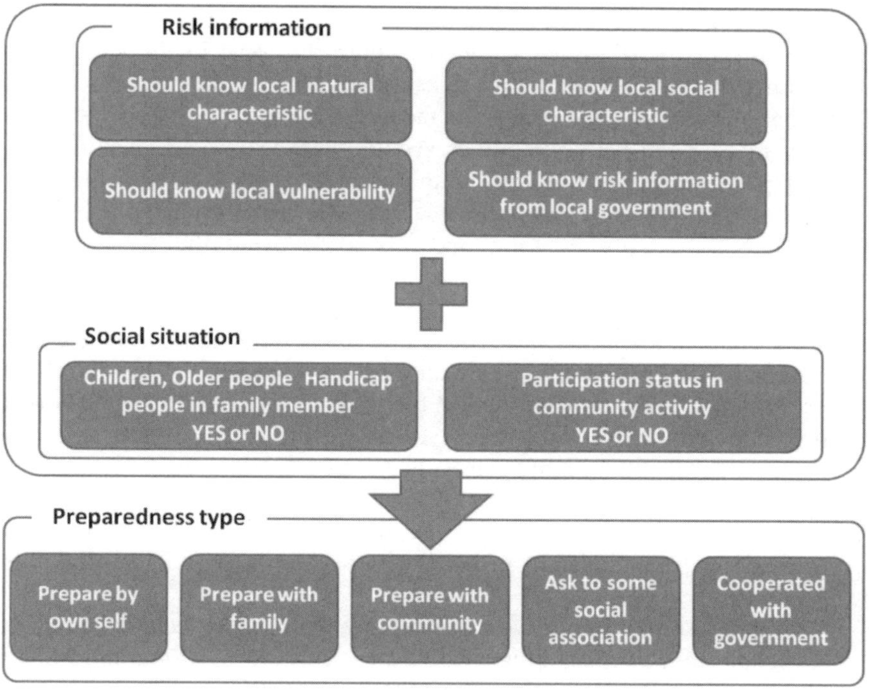

Fig. 6. Educational Needs and Types Based on Local Characteristics.

damages to their lives and livelihood. But some areas suffered only minimal damage because of Bio-Shield (e.g., mangroves and corals). Some parts of South India keep mangrove areas and were spared the full impact of the Indian Ocean tsunami. However, recently, mangrove areas have been converted to shrimp ponds and boat harbors for business. In developing countries, the government cannot pay much money for structural disaster countermeasures such as concrete walls. Therefore, Bio-Shield is an important countermeasure tool for disaster prevention. There is a need for people in coastal areas to understand the role of the mangroves and to learn to coexist with the mangroves while looking for other livelihood means. From the result of a questionnaire about prioritizing local community needs, it is clear that there are no specific needs on disasters (Ogino, Shaw, & Takeuchi, 2010). Most of the respondents would rather prioritize health, poverty reduction, and education. Therefore, it will be difficult to focus only on disaster risk communication. Other issues should be addressed as well.

This is possibly the standard case in most of the communities unless they are periodically hit by disasters. How to include the topic of disasters when addressing these aspects of life is the key point for sustainable and strategic risk communication. One way is to include development support such as improvement of self-help capabilities. Another way is to focus on education. School plays an important role in the social system. Teachers impart a lot of information (not only disaster education) to students in school. After school, students go home and talk about what they learned with their family. Families have a role in information dissemination and information can be spread quickly in the community (Takeuchi & Shaw, 2010). Elementary school especially is one of the vital infrastructures and landmarks in the community. Sometimes festivals are held in the school. At those times, the community gathers in the school. Community people have a relationship with the school teacher, and he/she is well respected, and regarded as local elite in most places. From such a viewpoint, the school is an important information center in the community. At the start, the information sender does not have arrows directly leading to the information receiver, but when a university or NGO approaches the school, new linkages will develop.

As in India, some communities and families have serious economic issues. Then people cannot focus only on disasters. Therefore, it is important to cooperate with other community problems. Disaster education must be sustainable. Cooperation of community issues and disaster education need to be effective, interesting, and sustainable.

The Case of Indonesia

The world's largest archipelago, Indonesia, is one of the most hazard-prone countries in the world in terms of tectonic activity. Situated on a major tectonic fault line which produces frequent natural disasters including earthquakes, volcanoes, and tsunamis, the country also faces frequent mudslides, landslides, and flooding. These disasters have a major effect on the population, with thousands of people killed and displaced each year, as well as a huge social and economic impact on the country. The national government recognizes the need for long-term investment in education, which effectively addresses the need of adopting a necessary culture of disaster prevention and mitigation, to help reduce loss of lives and properties from these hazards. One approach that has been taken is to integrate disaster risk reduction into the education system.

As part of the strategy to reduce disaster risks, *Muhammadiyah/Aisyiyah* Disaster Management Centre (MDMC), in cooperation with AusAID and with technical support from the Center for Disaster Mitigation-Institut Teknologi Bandung (ITB), has conducted a series of activities in order to enhance disaster preparedness among students and communities around *Muhammadiyah-Aisyiyah* schools. *Muhammadiyah* as one of the oldest civil society-religious-based organization in Indonesia has taken the initiative to commence disaster education. The activities aim to enhance knowledge, awareness, and preparedness among students, and are promoted at schools as well as in the community. The initiative's background was also underlined by the findings from a study conducted by Shaw, Shiwaku, Kobayashi, and Kobayashi (2004). It showed that while school-knowledge-based education programs are useful, additional activities (e.g., family and community education) were more important for converting knowledge into action. The findings suggested that a systematic and sustainable intervention between education in the schools and communities is essential to optimize disaster awareness and knowledge process, as well as rescue efforts to reduce disaster risks, which in turn develops the capacity in disseminating disaster education knowledge to others (family members, neighbors, etc.).

A comprehensive approach is then needed to combine social intervention at the community level that is supported by a change of policy and good governance at national and global levels (Child Disaster Awareness for Schools and Communities [CDASC], 2008). In Indonesia, Law No.24/2007, Presidential Regulation 28/2008 concerning *Badan Nasional Penanggulangan Bencana* (BNPB) as National Disaster Management Body, forms the legislative basis in promoting disaster education in the country. In response to the national government's call, *Muhammadiyah*, in partnership with the Australian Government through AusAID, designed a program, Child Disaster Awareness for Schools and Communities/CDASC (2008). This program enhances children's awareness to disaster, both at school and in the community to enhance community resilience, which is highly dependent on the capacity to prevent, to be prepared, and to respond to disaster threats. CDASC-MDMC objectives and targets were based on the assumption about student and youth position as agents of change. Following are the achievements of the activities of CDASC.

Development of Muhammadiyah *Capacity to Deliver School Disaster Awareness and Planning Program*

Student Working Sheets (LKS) have been developed to understand the awareness among students and teachers. Based on the data obtained, guidebooks and manuals on disaster education were developed. Assessment of the contents is still on-going. The schools are currently performing fairly well, however, without the support from *Muhammadiyah* boards at the district and sub-district level (particularly *DIKDASMEN*/Ministry of Education), school-based disaster education remains vulnerable. For the community-based activities, with the inclusion between *Muhammadiyah* youths and local youths, it has established the association as the forefront of sustainable awareness campaign by integrating other social issues. Cooperation with local government remains minimal and strengthening of such collaboration is strongly encouraged in the future. In general, the activities implemented in the schools and communities have enriched teachers, students, volunteers, and community members.

Development and Distribution of Disaster Education Materials in Various Media

During the period between 2006 and 2008, reading materials, games, evacuation route, animation film, semi-documentary film, documentation clips, and blog have been developed. These materials take into account the local context and use simple explanatory terms. Unfortunately, there is lack of community involvement in the development of such education materials. Village vulnerability maps and evacuation maps were one of the few activities that were well received by the community. *Kampong* newspaper was developed help disseminate local information within the community.

Community-Based Disaster Risk Management (CBDRM)

The youth groups agreed to continue facilitating the implementation of the community-based activities initiated by CDASC. Community organizations such as *FASB* in *Garut (West Java)*, *KSB* in *Padang (West Sumatera)*, *KASB* in *Rejang Lebong (Bengkulu)*, and *Club Siaga* in *Bantul (Yogyakarta)* became actively involved in these activities. Through the initiative by *Muhammadiyah*, the community has developed their social network and sense of responsibility in the respective community, which achieves the aim of building social capacity and enhancing community resilience. The efforts by *Muhammadiyah* have contributed significantly to the communities and is one of the classic examples of how disaster education is provided to the children, youth as well as the community by the support from civil society-religious-based organization in Indonesia. Through this process, there is a synergy between the *Muhammdiyah* school children and youths with

Muhammadiyah people/members and eventually with the communities surrounding. This, in turn, forms the social network that enhances disaster education among the people.

TOWARD SUSTAINABLE DISASTER EDUCATION

The role of family and community participation is crucial for the enhancement as well as the sustainability of disaster education. As described by Nimpuno (2007), in the immediate emergency period, assistance includes searching for and rescuing victims, and transporting them to nearby medical facilities. Due to the close proximity, family members provide assistance (e.g., emergency first aid, repair work, emotional support, etc.) within the family as well as to the neighbors in times of emergency. In order to be equipped with the skills to self-help and assist others, sufficient knowledge in disaster education is necessary, thus, the family is a strong element in disaster education framework. And in terms of coping mechanisms, the relationship between family and community through disaster education is inter-related and should not be overlooked. It is important to know how individuals, families, and communities in different societies adjust and cope with the disaster situations. The strengths of relationships, the concept of neighborliness, roles, and responsibilities, obligations, and reciprocities, as transmitted through the socialization process and life experiences play a significant role in advocating disaster education as coping mechanism. Additionally, the role of community in mobilizing the disaster education process through social networks is essential. These social networks have existed over generations (mentioned in the previous section) and may play a role before, during, and after disaster. The main strategy for disaster education is how to utilize and make effective the operation of these networks, because these have the potential to mobilize the disaster risk reduction actions in mass form.

The role of social mobilization in disaster reduction has been widely recognized in various parts of the world (Nimpuno, 2007; CDASC Final Report, 2008). Social mobilization is helpful in disaster reduction through community participation. The nature of the social organization and the adjustment strategies are related to the level of development. In relatively under-developed rural society, the coping pattern is usually characterized by an absorptive capacity (Nimpuno, 2007). The adjustment strategies of people can be boosted by external catalysts if they can appreciate people's needs and aspirations. Networks link people with shared needs, values, and aspirations. The key to disaster education is to mobilize communities to include vulnerability reduction in their development efforts. The work of housing cooperatives, school children, women's organization, youth, and religious groups should aspire to achieve development and reduction of the disaster risks. Social mobilization is no doubt a people's strategy in disaster education. Mobilization is a concerted attempt to organize people toward a common goal. Participation in achieving a common goal may be voluntary. The direction and forms of mobilization will depend upon who takes the lead in the mobilization of disaster education process. Mobilization may occur at different levels, i.e., family, community, national level, and the tasks and objectives of mobilization differ accordingly. Ultimately, the role of families and communities is important in disaster education; as the actors, as well as the key stakeholders, are linked by the mobilization process.

There are various ways in which disaster education is transferred within families and communities. As mentioned earlier in the chapter, implementation of disaster education can be school-based, women's activities in women organizations, youth groups, and community-religious-based. Mobilization plays and important role as well. The facilitators for disaster education in the family could be children, women, as well as senior members. They can be "transfer agents" within the family and between the family and community and/or school. Approaches and methods for conducting disaster education to families may be school-based (e.g., town watches) and/or activities organized by women organizations and groups. As for disaster education in the communities, the social network groups and "transfer agents" play an important role in communicating and transferring the knowledge (disaster education) for enhancing community resilience.

Disaster education should be woven and embedded in the daily life and form a culture of safety, whether in the instrumental acts of assessment, planning, and design of environmental protection and development of response capacity (Petal, 2009). The progress in disaster education will depend upon an iterative process that integrates scientific, technical, and indigenous knowledge, and experiential knowledge. Furthermore, information sharing, cooperation, and collaboration that promote good governance are essential. Learning to act locally as suggested by Petal (2009) is necessary to ensure a sustained understanding and action. It is important for families and communities to recognize their roles as important agents of disaster education.

Family and community hold an important role in the advancing of disaster education. In addition to the elements that mold the disaster education in family and community (i.e., custom/habit, way of life, religion, traditional and indigenous knowledge, and the nature of the community), there is still the challenge of how to adjust disaster education that is suitable to the family and community. Because all people are not equally vulnerable, similar disaster events may cause different consequences in different regions. These events could also be differed in their impacts on various groups within the society (population). Variability may occur at different levels and in general terms, vulnerability might differ as well, depending upon sex, age, ethnic group, class, etc. Due to differences in vulnerability, variation in behavior or disaster risk reduction actions might occur among different groups of people, including in families and communities. It means that differences in vulnerability across groups and classes lead to different coping strategies in the face of disaster. An understanding of the differences in coping strategy among family and community is thus crucial in devising disaster education. Involvement of other stakeholders, which exist within the society, such as organizations, communities, agencies, departments, jurisdictions, and policy-making bodiesmust recognize and perform their roles in a large cooperative effort toward the development of disaster-resilient communities.

ACKNOWLEDGMENTS

The authors are thankful to JSPS Bilateral Funding [Japan-India program], GCOE-ARS and MEXT Scholarship [to the second author] for undertaking this study.

REFERENCES

Child Disaster Awareness for Schools and Communities (CDASC) (2008). *Local initiative for global resilience*. Completion Report of 2008. CDASC Management as part of Muhammadiyah Disaster Management Centre, Central Board of Muhammadiyah, Jakarta.

Ensor, M.O. (2010). Education in emergencies: Lessons in child protection and disaster risk reduction. Available at https://www.conftool.com/Ensor-Education_in_Emergencies-255.pdf?Ensor-Education_in_Emergencies. Accessed on December 19, 2010.

Nimpuno, K. (2007). Participation in structural upgrading: A training course in structural upgrading for NGOs. Document No. 4266. Available at http://desastres.usac.edu.gt/documentos/pdf/eng/doc4266/. Accessed December 19, 2010.

Ogino, Y., Shaw, R., & Takeuchi, Y. (2010). Community perspective of mangrove protection and its implication to coastal zone management. In: R. Shaw & R. Krishnamurthy (Eds), *Pichavaram, South India, communities and coastal zone management* (pp. 113–125). Singapore: Research Publishing Services.

Petal, M. (2009). Education in disaster risk reduction. In: R. Shaw & R. R. Krishnamurthy (Eds), *Disaster management: Global challenges and local solutions* (pp. 285–320). India: Universities Press.

Shaw, R., Shiwaku, K., Kobayashi, H., & Kobayashi, M. (2004). Linking experience, education, perception and earthquake preparedness. *Disaster Prevention and Management*, *13*(1), 39–49.

Shiwaku, K. (2009). Essentials of school disaster education: Example from Kobe, Japan. In: R. Shaw & R. Krishnamurthy, R. (Eds), *Disaster management: Global challenges and local solutions* (pp. 321–387). India: Universities Press.

Takeuchi, Y., & Shaw, R. (2009). Gender dimensions in risk communication a perspective from a sediment disaster in Hiroshima, Japan, regional development dialogue (RDD). *United Nations Center for Regional Development*, *30*(1), 63–75.

Takeuchi, Y., & Shaw, R. (2010). Risk communication strategy for effective coastal zone management. In: R. Shaw & R. Krishnamurthy (Eds), *Communities and Coastal Zone Management* (pp. 155–164). Singapore: Research Publishing Services.

Victoria, L. P. (2009). Community capacity and disaster resilience. In: R. Shaw & R. R. Krishnamurthy (Eds), *Disaster management: Global challenges and local solutions* (pp. 338–351). India: Universities Press.

CHAPTER 5

ESSENTIALS OF HIGHER EDUCATION IN DISASTER RISK REDUCTION: PROSPECTS AND CHALLENGES

Rajib Shaw, Fuad Mallick and Yukiko Takeuchi

INTRODUCTION

When discussing disaster education, the usual focus in more on the school or family or community education. Very little focus has been given so far to higher education. However, higher education (college and university) is the key to professional development in the subject. Higher education in disasters is still lacking in most countries and regions. In this context, the lessons of environment or the field of sustainable development can provide useful tips. Of equal importance to higher education is not only the curriculum, but the approach or mode of delivery. To develop an appropriate higher education, a system of educational governance is important (COE, 2005). Given the role education has for overall societal and economic development, it is necessary to ensure the responsiveness of higher education to the changing needs and expectations of society. In this respect, it is important to ensure participation of external actors in the governance of higher education and to allow the flexibility to accommodate the continually change needs and requirements over time. COE (2005) made several recommendations for

Disaster Education
Community, Environment and Disaster Risk Management, Volume 7, 95–113

ISSN: 2040-7262/doi:10.1108/S2040-7262(2011)0000007011

higher education that can be considered as the base of disaster education in colleges and universities. These include: serving the needs and expectations of the society, having appropriate academic freedom, having a process of setting up long-terms goals and developing appropriate strategies for achieving them, providing reasonable scope of innovation and flexibility in research, promoting good educational governance through regional and international networks, and ensuring quality control of teachers and students.

For development to be sustainable, mitigation of the destructive effects of natural disasters remains a challenge. Proper application of scientific and technical knowledge on disasters offers an opportunity to reduce the vulnerabilities and risks of various social groups. Disaster education is now recognized as an essential element in formulating the appropriate disaster risk reduction strategies for any country. Over the last few decades, there has been remarkable progress in developing the theoretical basis for disaster management. The advancement in science and technology allows better understanding and presentation of risk and vulnerability issues; application of remote sensing techniques and GIS has significantly contributed in this regard. Social scientists have been successful in exploring the various social dimensions associated with disaster risks and vulnerabilities.

Higher education in disaster risk reduction (DRR) is a multi-disciplinary issue. It encompasses all faculties of knowledge ranging from science, social science, humanities, and so on. It has not been long that some formal academic degrees have been offered in the field of DRR from a few academic institutions worldwide. Since long before offering academic degrees, many institutions around the world have been conducting disaster-related research and offering training programs of varying duration. The objective of these research and training and degree programs are mainly to foster local or regional needs.

It is important to facilitate DRR education at tertiary levels through development of a regionally suitable and sufficiently flexible curriculum structure. Before embarking on developing a curriculum structure for DRR education, particularly at the post-graduate level, it is important to differentiate DRR education from DRR training. This is because prior to the evolution of mainstream DRR education, DRR training either in the form of response and recovery or preparedness was fairly widespread across agencies concerned with it. DRR education is not only about the creation of well-versed professionals in DRR but is also a vehicle for knowledge accumulation and, importantly, knowledge creation. While there are academic programs that deal with disasters from the perspective of the disciplines that hosts them, i.e., geology, engineering, geography to name a few, DRR has yet to visibly emerge as a discipline on its own, which takes comprehensive account of all components.

With the above background, this chapter focuses on the importance, evolution, and process of higher education in the disaster field. The chapter consists of five sections. The first one describes the evolution and lessons from the higher education in sustainable development. The next section highlights the emerging issues of higher education, followed by networking in higher education in DRR, networking in higher education in climate change adaptation, and future directions of higher education.

HIGHER EDUCATION FOR SUSTAINABLE DEVELOPMENT (ESD)

CRE (1995) developed the university charter for sustainable development in 1995. The charter points out that education at all levels, especially university education for the training of decision-makers and teachers, should be oriented toward sustainable development and foster environmentally aware attitudes, skills, and behavior patterns, as well as a sense of ethical responsibility. Education must become environmental education in the fullest sense of the term. To conduct appropriate education in sustainable development, the charter urges the universities to take appropriate actions such as: institutional commitment to the principles and practices of environmental protection; promoting environmental ethics among teachers and students; education of university employees; programs in environmental education, inter-disciplinarily; dissemination of knowledge; networking; partnerships; and technology transfer. The point most applicable to higher education (HE) was the need for the reorientation of existing education to take sustainable development principles into account (Brunton, 2006). In addressing this point, McKeown (2002) makes the argument that countries should be careful not to confuse reorientation and increasing the amount of education. Despite numerous bodies arguing that economic, social, and environmental issues need to be equally weighted, it is often the environmental angle that comes to the fore. "Greening the university" or "greening the curriculum" have become commonly used phrases that tend to refer to the integration of environmental perspectives into university operations and teaching (Alabaster & Blair, 1996). Brunton (2006) from his study concludes that the ingredients for effectively incorporating ESD into teaching and learning practices include:

- Full integration of ESD into the curriculum (i.e., not a one-off or final-year option module), with continuity (i.e., not a three-year project that ends in a report)
- Student-centered activities and assessments that reward critical thinking and reflective learning, e.g., use of problem-based learning, projects, case studies, portfolios, field studies
- Trans-disciplinary teaching, with modules that are taught by staff from a range of disciplines and encourage contact between students from different subject areas
- Teaching that emphasizes that ESD is an ongoing process, i.e., part of a lifelong learning journey where answers are not hard and fast.

In an interesting article, Fortuin and Bush (2010) mentioned that to understand the complex issues environment and sustainable development, the scientists need "boundary- crossing skills" next to domain-specific knowledge and communicative and social skills. Citing the example of the European Workshop (EUW) as an interdisciplinary course, the paper suggests to incorporate course structure [stage-based with specific phases focusing on specific inter-disciplinary topics], course components [including a matrix approach with course reporting and fieldwork], communication among students, and role of teachers as essential to facilitate the "boundary-crossing skills" among the students. Two components of the EUW, the matrix approach and the field work, particularly contributed to enhancing students' awareness of disciplinary and cultural boundaries. They also contributed to the students' appreciation of using different disciplinary and cultural perspectives in solving problems. The students developed positive attitudes or habits to crossing boundaries, a precondition for being able to cross them. By working on a real project in an intercultural setting, the students can be confronted with shortcomings of scientific research and the often politicized nature of environmental management. Learning to cope with these issues by questioning the reliability of information and realizing that decisions are often made in a particular context, exposed the students to the central challenges of crossing boundaries between theory and practice, disciplines and cultures. This realization can be transferred into research and professional skills as they advance with their academic and professional careers and will be further exposed to the complexity of environmental and societal problems. Realizing that one should cross boundaries to solve problems could be one of the most important elements in their education (Fortuin & Bush, 2010).

There exist several other models of deep learning (Warburton, 2003), sustainability trans-disciplinary education model (STEM) (Clark & Button, 2011), and competency enhancement for interdisciplinary issues (Parker, 2010) in higher education. Deep learning is a key strategy by which the students extract meaning and understanding from course materials and experiences (Warburton, 2003). Because of the range and interconnectedness of environmental, social, and economic issues, and the importance of interdisciplinary thinking and holistic insight, deep learning is particularly relevant in the context of education for sustainability. However, deep learning can be inhibited if the existing interests or backgrounds of students have a strong disciplinary focus. The STEM integrates the sciences, arts, and aesthetics, and the university with the greater New Britain community, and beyond (Clark & Button, 2011). Academic areas include geography, environmental science, communication, art history, aesthetics, and teacher education. The trans-disciplinary methodology was integrated in a learner-centered design. As a result of the mutual learning implicit in the STEM, all participants expanded each other's understandings of sustainability. Students were learning from instructors, instructors were learning from students, students were learning from students, instructors were learning from instructors, and all were learning and sharing knowledge with the greater community. As a result, all participants gained a deeper and broader understanding about human-environment relationships and how humans impact natural resources. Competencies are at the heart of the new forms of inter-agency and inter-professional working that is increasingly recognized as essential to deliver care and sustainability in a joined-up world (Parker, 2010). The literature review demonstrates developments toward action competencies in inter-disciplinary for sustainability but with an over-reliance on students guiding their own practice and reflection. Findings highlight potential elements of a more widely informed knowledge literacy, including philosophical, socio-logical, and cultural aspects, that is needed to support the development of these competencies.

In a review of the progress of higher education in sustainable development, Ryan, Tilbury, Corcoran, Abe, and Nomura (2010) have pointed out that the Asia-Pacific region offers many creative initiatives and shows considerable progress in ESD and in understanding the learning dimensions of sustainability. At the same time, it mirrors global trends that further work is needed to promote systemic change in educational arenas, particularly in terms of strategic integration within higher education institutions. Nomura and Abe (2010) pointed out that encouraging leadership development for sustainability among university executive staff members is critical to continuing and strengthening efforts in this area in Japanese higher education. In contrast, the example from India shows that several principles of sustainable development are embedded in the country's education policy (Chhokar, 2010). It is perhaps the only country where the highest court has mandated environmental education at all levels of formal education, which includes a compulsory undergraduate course. However, the challenges of implementing this requirement effectively are hampered by lack of inter-disciplinary competence among staff and students, and traditional methods of assessment in higher education. In the case of China, teaching about sustainable development has been integrated into technical fields, especially at universities in major cities, and research has been undertaken to develop economically effective and environmentally friendly innovations (Niu, Jiang, & Li, 2010). However, resources and capacities are still very limited, given the ambitions associated with sustainable development in China. Different examples demonstrate the need to harness national policy, to develop local and regional initiatives, and to work effectively toward more profound change in higher education curricula and through collaboration with external communities and stakeholders (Ryan et al., 2010).

EMERGING ISSUES OF HIGHER EDUCATION IN DISASTER RISK REDUCTION

Disaster risk reduction (DRR), an inter-disciplinary subject, has links to other sectors such as environment, development, and human security. The link between environment and disaster is prominent in the area where natural and social issues merge, and this is specifically prominent in the rural areas where most of the communities depend on agriculture and natural resources for their livelihood. These issues are linked with the overall concept of human security (Shaw, 2006). Climate change impacts are often regarded as the missing link between environment and disaster. However, the relationship is not clearly reflected in the national policies and international and local actions. The following section shows some of the key issues to be reflected in higher education in DRR.

Environment Disaster Linkages

The link between development, environment, and disasters is a very deeply entrenched one. Unplanned, ad hoc, and poor development is directly responsible for a significant part of the vulnerabilities observed in the region. Large-scale industrial developments unmindful of related risks and pollution considerations, the rise of high-density settlements with inadequate infrastructure, non-engineered buildings, all have contributed to high levels of risk. The intermediary in this process is often the environment, as can be seen clearly in the case of climate change that has been established to have arisen out of anthropologic causes. While there is very popular recent focus on carbon footprints, the concept of ecologic footprints has existed for a long time in the academic domain. The fact that the environment has a limited carrying capacity beyond which it cannot support consumption and emission levels has been a well-known fact for very long, and has been articulated in scientific terms for many decades. Yet ecologic footprints have gone on increasing exponentially with increasing consumption patterns that have accompanied development and economic growth.

Hydro-meteorologic Disasters

The most significant increasing trend that threatens vulnerable populations in the mountain regions, river basins, arid swathes and coastal stretches of Asia is one of increasing hydro-meteorologic disasters. In terms of climate change-induced catastrophic events, these threaten people with shocks such as cyclones, cloudbursts, flash floods, and urban floods. On the other hand, in terms of prolonged stresses, they threaten with drought and water stresses. Their impact on urban settlements is also increasing, though not as visible and noticeableas the case of catastrophic events. Unmanageable migration, often in waves or migration storms that are related to distress migration, are burdening the already bursting Asian cities to a point of breakdown. Urban droughts are an intimidating phenomenon that looms in the face of many Asian cities already reeling under severe and prolonged water stresses. What climate change threatens to do to the water towers, the mighty glaciers, and water reservoirs in the mountain regions only compounds the seriousness of these threats.

CCA-DRR Synergy

Climate change adaptation (CCA) and disaster risk reduction (DRR), though broadly understood to be linked in some ways, have not yet been taken as a holistically linked complementary set of actions that require collaborative and coordinated action by all concerned stakeholders. The significance of CCA-DRR synergy cannot be felt more by vulnerable communities who do not feel the impact of climate change or natural disaster sectorally, but it hits them as a combined whole with devastating effects. It needs to be appreciated that a piece-meal, sectorally split approach to this complex set of problems will not bear fruit. Recent work by some of the Asian University Network of Environment and Disaster Management (AUEDM) university partners has shed light on the intricate linkages between cross-sectoral development activities, their impact on the environment, subsequent detrimental impacts of a deteriorating environment on human life, and the integrated approach needed to address this combined threat of climate change and disasters. Such an understanding can be very meaningfully deployed at various levels – from governance to voluntary action to education, and can go a long way in developing community-based and environment-based resilience to climate change as well as disasters.

NETWORKING OF HIGHER EDUCATION IN DISASTER RISK REDUCTION

AUEDM is a unique initiative of prominent Asian universities that come together to share knowledge resources related to environment and disaster risk management among themselves and with the larger group of stakeholders working on these issues, in addition to conventional national and thematic boundaries. AUEDM members work in close collaboration to conduct education and research, share findings, and find ways forward in a region that is increasingly threatened by climate change impacts. AUEDM also works closely with governments, international agencies, and corporate and civil society organizations to establish collaborations that eventually lead to resilient communities. AUEDM reflects each member's commitment to implementation-oriented education and research in the field of environment and disaster risk reduction. The objectives of AUEDM are:

1. To share and work together (bilaterally or multilaterally) in promoting environment and disaster risk reduction in higher education (focusing on, but not restricted to, post-graduate education)

2. To collaborate on field-based and policy-oriented research focusing on different aspects of disaster risk reduction and environmental management
3. To broaden the scope of education and learning in the environment and disaster risk reduction field through collaboration with diverse stakeholders including NGOs and local governments
4. To document, develop, and disseminate knowledge products in the field of environment and disaster risk reduction
5. To provide a forum for consultation, information sharing, and cooperation among universities on matters and themes of common interest
6. To enhance recognition of the vital role of universities in implementation-oriented education and research in environment and disaster risk reduction

One of the specific features of the AUEDM is close cooperation with the civil society organizations. Non-government organizations (NGOs) have direct field access, and experiences in grass-root project implementation. However, these experiences are not properly reflected in the educational curriculum. Thus, the network aims at bridging academic research, education and field practice. Some of the highlights of the university-NGO cooperation are:

1. Quality of knowledge and information: All participating universities in the targeted countries are esteemed organizations in the field of disaster risk management. Therefore, it brings high quality knowledge and information.
2. Extensive network: The four universities have the largest networks in the tsunami-affected areas, and thus ensure that the knowledge product will have largest circulation in future.
3. Ensuring sustainability: Through development of the certificate courses and customized courses, young professional development will be ensured, which is linked to the sustainability of the disaster preparedness activities in the targeted countries and communities.

AUEDM Started from An Imperative

AUEDM was conceived and pursued by its member organizations based on this common understanding and motivation. It has come about from needs that appear to be crucial for the survival of millions of poor and vulnerable men, women, and children living on the margins of society in Asia. AUEDM come together for reasons of educational, research, and networking imperatives.

Educational imperative: To discuss the status and scope of environment and/or disaster risk reduction curriculum in the higher studies in each university

Each country has its own perspective. Some countries have a full 2-year DRR master's program. Some universities have some modules of DRR in the postgraduate programs. Therefore, the attempt is not to standardize the program, but to learn and understand the process in DRR. The challenge is how effective the process can be customized into each context.

Research Imperative: To Discuss the Possibility of Climate Change Adaptation as the Key Entry Point of Collaborative Research

Each country has a high prevalence of impacts of climate change being borne by the most vulnerable communities. Impacts are most visible on coastal, mountain, urban poor, and migrant communities. Because adaptation is a relatively new subject, heavy investments need to be made in research on effective local adaptation as a means for coping with imminent climate change impacts and linked disasters.

Network Imperative: To Discuss the Establishment of the Asian Universities Network

While there are integral commonalities in the vulnerability context and the nature of impacts, the local setting and contextual nuances are highly varied across Asian countries. Networking is the only way to share knowledge and experiences, and to draw lessons based on principles derived from practices. The network is thus expected to go a long way in the development of a regional knowledge base, making it accessible for practitioners, and using it to influence the policy environment.

Skeletal Base for Disaster Education in AUEDM

Considering the multitude of issues associated with DRR and to address the differential dimensions of hazards in different geographic contexts, the curriculum should be "adaptable and contextualizable." The skeletal base for DM curriculum can be thought with four different heads namely: the

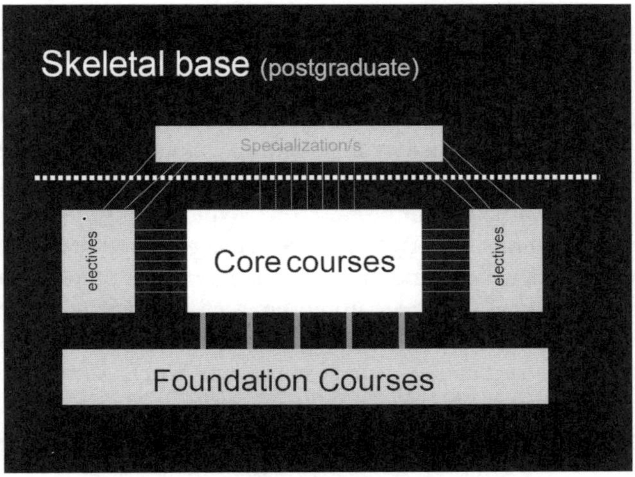

Fig. 1. Structure of Disaster Education Course in AUEDM.

Foundations, the Cores, the Electives and the Specialization with a view to fostering individual student's needs and capacities (Fig. 1).

In the **foundation** course, lessons on the origin and dynamics of hazards and disasters will be given. Besides the scientific aspects, the social issues in the construct of disaster, as well as the cross-disciplinary dimensions of are to be covered. The **core** courses will deal with imparting in-depth knowledge for the understanding of risk and vulnerabilities issues across all phases of disaster management. Application of various tools and techniques for assessment and analysis of risk and vulnerabilities in pre-disaster situation, as well as for the need assessment in post-disaster situation, must be discussed under the core courses. **Elective** course will be designed with a view to fostering individual interest where students can choose from a range of choices. This will consequently help students decide on their specialization where they carry out a dissertation work. Based on the academic and professional backgrounds as well as the aspiration for a future career, students will decide on the areas of their **specialization**. Based on the nature of the university, the courses can be as follow:

- PG Diploma: 6 to 8 months
- Master's: 2 years
- M.Phil: 1 year after Master's; 3 years after Bachelor
- Ph.D: 2 years after M.Phil; 3 years after Master's

Mode of Delivery of Higher Education in DRR

Lectures and Coursework

Lectures are probably the traditional teaching method in many circumstances and for many students, especially for communicating conceptual knowledge, and where there is a significant knowledge gap between lecturer and audience. In DRR the lectures are effective because they exploit the spontaneous human aptitude and communications. Experienced teachers should be invited to deliver lectures to the students of disaster management. Teachers from universities, disaster or environment ministries, fire stations, life-line related companies, city governments, volunteer organizations, and so on should be involved. In some cases there may be students who have experience in DRR in which case the lectures can be supported by interactive sessions for cross learning.

Fieldwork

Fieldwork is the means of gaining first-hand knowledge of an environment. This experience plays a fundamental role in the teaching of disaster risk reduction. This activity makes the lesson more interesting to students and through it students acquire certain skills, which are rarely learnt in the classroom. Some of these skills, which they can use in everyday life, include the skill of observation, of experimentation and investigation, problem solving, cooperation, communication, and decision-making. In DRR the students need to carry out data collection and calculations in order to achieve results and they are also encouraged to use different research tools. Tools such as questionnaires, interview schedules, FGDs (focus group discussions), etc. are common research tools. For students, whether they are in Master's level or in PhD, field work is to be compulsory. The prime focus of the AUEDM is to present an opportunity for graduates from a variety of disciplines to expand their viewpoint through a combined program of taught and research components. The field work/research will create deeper insight of the issues associated with risk and vulnerability assessment, preparedness, response, recovery, reconstruction and rehabilitation, development, and the enhancement of community capacity in a broad range of disaster scenarios.

Internship

Long-term (from 3 to 5 months, depending on the nature of the course) internship programs for the students can be considered as a compulsory subject in DRR education in the university. The internship needs to be long-term, and through this education process, the students need to go out of the university boundaries. The internship places can be research institutions, government departments (national and local), non-government organizations (NGOs), international organizations and bilateral donors and private sectors. Specific monitoring mechanism needs to be placed to ensure the quality of internship. The internship program provides practical, real-world knowledge to the students. It also helps them in future job prospects.

Research Linkages
Academic institutions/universities and aid agencies have significant contributions toward sustainable DM. School, college, and university programs play a key role in establishing successful partnerships with operational organizations because of the technical content of these areas and the expectation of outreach among these departments. Universities have provided research opportunities and contextual knowledge that are of potentially great value for operational agencies. Increasingly, university settings may serve operational agency staffs that are prepared to reflect, write, and publish on the insights they have gained from their experiences. Academics, for their part, already derive considerable advantage when operational agencies facilitate their access to the field. In both applied and social sciences, essential field data may be difficult or impossible (or dangerous) to gather without an affiliation that is accepted and locally trusted. It is important both for that organization's employees and university faculty members to have the opportunities to spend time in each other's programs.

Market Demand/Socialization
To some extent, a disaster is predictable, but there are varying degrees of uncertainty as to how and when it will occur. The incidence of a disaster creates varying degrees of anarchy combined with a variance between resources and needs. Therefore, in order to reinstate an affected society back to its pre-event position, extraordinary efforts are required. While, in the past, responses were slow, today we can rely on the provision of timely relief, but the precision of what we provide to meet the needs of the affected community leaves much to be desired. There is a need of expert people in the field for disaster management. This need can be socialized through advocacy events and effective communication with the relevant institutions, departments, and organizations. Disaster risk reduction should be mainstreamed with all the developmental activities. There should be a separate section in each department and ministry for disaster management. These qualified graduates should be placed in departments and organization such as authorities at various levels, emergency response and coordination bodies, early warning, flood control and disaster monitoring bodies, universities and research institutes, non-government

organizations, community-based organizations, and international development organizations, local and municipal administrations, with civil defense, media or hospital management, training institutions, consulting companies, or private business. These graduates have high capacities to adjust to the steadily rising complexity of an interlinked and globalizing world because of the interdisciplinary character of the education.

HIGHER EDUCATION IN CLIMATE CHANGE ADAPTATION

The United Nations University has started an innovative networking among universities called UN-CECAR. The University Network for Climate and Ecosystems Change Adaptation Research (UN CECAR, 2009) was established in 2009 by leading universities in the Asia Pacific region to strengthen the higher education sector to respond effectively to climate and ecosystem change. Research and education are the main focus of UN-CECAR. The network will bring together all-available resources and expertise across disciplinary lines to work collaboratively to enhance understanding on how climate change is affecting roughly two-thirds of the world's population who live in Asia Pacific region, and advance adaptation research for the design of appropriate policy and development strategies. The network was conceived out of a round of discussions held during a conference and a 2-day workshop on the "Role of Higher Education in Adapting to Climate Change" from June 10–12 in Tokyo, organized by United Nations University Institute for Sustainability and Peace (UNU-ISP) and Integrated Research System for Sustainability Science (IR3 S). The overall purpose of UN-CECAR is to reduce the Asia-Pacific region's overall vulnerability to the impacts of climate change, and increase local adaptive capacities through the leadership of higher education sector. UN-CECAR is targeting higher education institutions as it provides the ultimate source of human capital and knowledge for tomorrow's climate-risk resilient societies. The objectives are to collect the knowledge on climate change adaptation accumulated at the international level, and synthesize and customize it to the local level through the following:

- Assess existing and emerging climate change-related research and degree programs in the region;
- Identify areas of most need according to students and educators, while also ensuring that market demands for climate-change graduates are also being met;

- Initiate and support the development of educational programs comprising of: joint or dual degree programs common courses open for credit sharing schemes initiate and support the development of joint research programs to promote: resource sharing (experimental fields and facilities, modeling and forecasting systems, short-term training, available funds, etc.) frameworks and mechanisms for joint research programs, existing and proposed initiatives, fundraising activities, etc. initiate and support training programs that re-train professionals, specialists, policymakers, educators on emerging climate change issues, science, and methodologies;
- Create direct links with, and contribute to, existing and emerging international and national processes and international conventions, such as the work of the IPCC, the Bali Roadmap and Nairobi Action Plan, and the various national adaptation plans occurring in each country; create a framework for a regional curricula on climate change adaptation.

In the UN-CECAR, the curriculum development part consists of core modules, specialized modules and modules of cross-cutting issues. The structure of each course (science of climate and ecosystems, adaptation and mitigation, impacts and vulnerabilities) shall contain: core modules, specialized modules, and modules that cut across other disciplines. **Core modules** are generic principles for a particular focus area that may include climate change science, impacts and vulnerabilities, mitigation and adaptation, climate-related services (policy, law, institutions, economics, instruments), and decision-making, etc. Specialized modules are offered to allow students to study an issue in greater detail, and practice or apply it in the real world. This may include climate modeling, impact studies, vulnerability assessment, development of scenarios (based on specific local context), resource optimization, etc., as well as focused sector-specific modules to address issues in water, agricultural, infrastructure, biodiversity, and humanitarian assistance sectors. Modules with cross-cutting issues are those the subjects that are not limited to the area of focus, but will have a significant importance in the context of climate change. Examples of cross-cutting modules may include: systems approaches and integration (including working with circularity), decision-making under uncertainty (considering that risk is no longer what we are dealing with but uncertainty, which is beyond the realm of current risks), resilience (adaptive to governance, social-ecologic values), etc. Users such as university faculties and students will be able to select the appropriate modules that not only meet their needs, but provide the most fulfilling courses for them.

THE WAY FORWARD: GUIDING PRINCIPLES OF HIGHER EDUCATION IN DRR

As explained in the earlier section, disaster risk reduction (DRR) is an interdisciplinary subject, and needs to be practical and field oriented with a balanced approach of theory and practice. In reality, DRR higher education is practiced in different departments in the universities, starting from engineering, science, architecture, agriculture, economics, social science, and humanities. A comprehensive DRR course is desirable but possible not a practical solution in many countries, depending on the market mechanism of required professionals. DRR higher education can learn significantly from the process and approaches of sustainable development. The following sections suggest some future directions of higher education in DRR.

Inclusive Curriculum

Disaster type, nature, intensity, density, frequency, perception, damages, response, relief, recovery, prevention, mitigation, and preparedness vary not only in terms of location but also relate to the socio-economic conditions, technical capabilities, political priorities, and development agenda of particular societies. While each has its own priority, a curriculum structure needs to address issues at a general level that are inclusive of all for a well-rounded foundation to proceeded to specific issues. This requires a faculty network with technical support of the universities concerned in order to design a universally acceptable curriculum on DRR and climate change adaptation. This should consider the potential time investment of both students and teachers and optimize the opportunity for quality student faculty interaction in support of learning and sharing information on approaches and innovation in the field of DRR.

Theoretical Focus

The curriculum will focus on imparting education primarily in the field of DRR with climate change adaptation, DRR/prevention/mitigation as important components. The theoretical focus will be on the basic concepts and theories of DRR, climate change adaptation, global warming, and scientific understanding of the various types of disasters.

Field Orientation

The curricula will not only focus on theoretical knowledge but the faculty and students would undertake research on disaster-related issues. Exposure of students to real-life situations to assess vulnerabilities, mitigation, and preparedness measures will help them bridge the theories with the practice. Scientific approaches could be adopted to conduct research work on Disaster risk reduction more at the field level to make it valid, applicable, and authentic. New techniques could be identified and adopted in this regard particularly for risk, vulnerability, and capacity assessment. Tools from the relevant field could be merged in the field of DRR for the conduction of various studies and researches.

Multidisciplinary Approach

Disaster preparedness and management are multidisciplinary in nature. Various subjects like geography, environmental sciences, geology, economics, sociology, social work, psychology, medical sciences, civil engineering, city and regional planning, architecture, urban and regional planning, agriculture, forestry, animal/plant sciences and management sciences contribute to the field of disaster management. Keeping in view the DRR cycle, there are special areas of interest and research. The curriculum of DRR will focus on all subjects. This multidisciplinary approach will provide a vast field of research and coordination on one hand and also increase the need of understanding of the subject as a whole on the other. The important point is to streamline these multidisciplinary course topics and research to an integrated and coherent subject of disaster management.

Skill Enhancement

The curricula will focus on producing trained manpower. The training should be based on experiences learned from the previous case studies according to market demand. The trained manpower produced from the university would be engaged in the government/non-government/private sector institutions dealing with disaster preparedness and management and thus play an important role in minimizing the losses caused by the disaster through better preparation and management.

ACKNOWLEDGMENTS

The first author acknowledges the support from the Mitsui Foundation to set up the AUEDM and conduct higher education processes in DRR. The authors acknowledge the discussion with AUEDM members.

REFERENCES

Alabaster, T., & Blair, D. (1996). Greening the university. In: J. Huckle & S. Sterling (Eds), *Education for sustainability*. Oxford: Earthscan.

Brunton, K. (2006). Education for sustainable development: Principles for curriculum development in business subject areas. *Investigations in University Teaching and Learning*, *3*(2), 36–46.

Chhokar, K. B. (2010). Higher education and curriculum innovation for sustainable development in India. *International Journal of Sustainability in Higher Education*, *11*(2), 141–152.

Clark, B., & Button, C. (2011). Sustainability, trans-disciplinary education model: Interface of arts, science and community (STEM). *International Journal of Sustainability in Higher Education*, *12*(1), 41–54.

COE (2005). Council of Europe: Higher education governance between democratic culture, academic aspirations and market forces: Considerations and recommendations. Available at: http://www.coe.int/t/dg4/highereducation/governance/GOV_recommendations_EN.pdf. Accessed on February 2011.

CRE (1995). Conference of European rectors: The University charter for sustainable development. Available at http://www.iisd.org/educate/declarat/coper.htm. Accessed on February 9, 2011.

Fortuin, K. P. J., & Bush, S. R. (2010). Educating students to cross boundaries between disciplines and cultures and between theory and practice. *International Journal of Sustainability in Higher Education*, *11*(1), 19–35.

McKeown, R. (2002). Education for Sustainable Development Toolkit V2 [www]. Available at http://www.esdtoolkit.org/default.htm. Accessed on February 9, 2011.

Niu, D., Jiang, D., & Li, F. (2010). Higher education for sustainable development in China. *International Journal of Sustainability in Higher Education*, *11*(2), 153–162.

Nomura, K., & Abe, O. (2010). Higher education for sustainable development in Japan: Policy and progress. *International Journal of Sustainability in Higher Education*, *11*(2), 120–129.

Parker, J. (2010). Competencies for inter-disciplinarity in higher education. *International Journal of Sustainability in Higher Education*, *11*(4), 325–338.

Ryan, A., Tilbury, D., Corcoran, P. B., Abe, O., & Nomura, K. (2010). sustainability in higher education in the Asia-Pacific: Developments, challenges and prospects. *International Journal of Sustainability in Higher Education*, *11*(2), 106–119.

Shaw, R. (2006). Community based climate change adaptation in Vietnam: Inter-linkage of environment, disaster and human security. In: S. Sonak (Ed.), *Multiple dimensions of global environmental changes* (pp. 521–547). India: TERI publication.

UN CECAR (2009). University network for climate and ecosystems change adaptation research. Available at http://cecar.unu.edu/. Accessed on February 9, 2011.

Warburton, K. (2003). Deep learning and education for sustainability. *International Journal of Sustainability in Higher Education*, *4*(1), 44–56.

CHAPTER 6

INNOVATIVE APPROACHES IN DISASTER EDUCATION

Koichi Shiwaku and Glenn Fernandez

INTRODUCTION

In the previous chapters, disaster education was discussed based on the aspect of the place where disaster education was conducted – in school, in the household, and in the community. Generally, school disaster education is regarded as formal disaster education, while household and community disaster education as informal disaster education. School-based stand-alone courses are perhaps the easiest programs to implement on a large scale and within a short time frame (Petal, 2009). However, to achieve community-based disaster management, stakeholder involvement is important. If students learn with the community, the learning may be regarded as informal or semi-formal education. When the community, including students, learn about disaster management, local contexts are important to be considered. Anticipated hazards, stakeholders, availability of human and physical resources, extent of threatened or affected area, culture, history, and other various factors can be taken into account. In addition, when the community and students learn together, there are at least two actors. This means that disaster education programs should focus on both community members and students as the target learners. Therefore, such education programs cannot be discussed from the aspect of place of education like school, community, or household. Shaw and Takeuchi (2008)

Disaster Education
Community, Environment and Disaster Risk Management, Volume 7, 115–136

ISSN: 2040-7262/doi:10.1108/S2040-7262(2011)0000007012

emphasized the importance of the participatory approach. Thus, it is necessary to consider how education programs should be conducted. As described before, it is necessary to consider various factors and situations in order to provide disaster education programs that meet local contexts. In this regard, a standardized disaster education program is not appropriate. Therefore, people/organizations who/that organize disaster education programs should need to play important roles so that the disaster education program becomes effective. In other words, the organizers need to consider how they provide disaster education as well as what they provide.

To achieve disaster education discussed above, capacity building of organizers or facilitators is important. Shiwaku, Shaw, Kandel Chandra, Shrestha, and Dixit (2006) also pointed out that school teachers are important change agents. If the organizers or facilitators of disaster education have knowledge of disaster and disaster management, they can provide disaster education to some extent. But the disaster education can be more effective if they take appropriate approaches. In this case, enhancing the capability or capacity of organizers or facilitators to deliver disaster education is more important than increasing their amount of knowledge. Therefore, it is necessary to raise people with such capacity in order to provide more effective disaster education in more schools or communities.

Recently, information and communication technology has developed rapidly. The Internet is one of the examples. DesInventar (no date) provides hazard and disaster information. UNISDR (2007a) has PreventionWeb to provide various types of information on disaster management. These days, people can access information on disasters and disaster management easily through the Internet. But it is significant to find out what is useful information. But what is more important is that people not only see information but also actually use it. For example, many organizations are providing information on disaster education on their websites. But school teachers, community leaders, or other organizers/facilitators of disaster education are requested to apply disaster education as shown in the Internet. Utilizing information needs people to have appropriate ability.

This chapter consists of three parts. The first part is school–community linkage. How the school and the community can be involved in disaster education is discussed. The second part is about efforts for the promotion of disaster education. This part focuses on school teachers and community leaders so that they can provide disaster education effectively. The third part discusses the sharing of good practices on the Internet. Several websites are chosen as examples of information sources of disaster education and disaster management.

SCHOOL–COMMUNITY LINKAGE

This section introduces three cases of disaster education related to school–community linkage. The first case is learning from using local resources in Japan. The second case is from Nepal on community-based disaster management through school safety, which has been conducted by an NGO. The third case is the learning among school students and community in Japan.

Utilization of Local Resources: Case of Japan

The Environment and Disaster Mitigation Course of Maiko High School, a school located in Kobe, is providing a characteristic disaster education, as mentioned in Chapter 3. The education programs are conducted not only inside the school but also outside the school. This section is about utilizing local resources in the education process. Hyogo Prefecture, which includes Kobe City, is the area affected by the Great Hanshin-Awaji Earthquake. In 2002, Hyogo Prefecture established "The Great Hanshin-Awaji Earthquake Memorial – Disaster Reduction and Human Renovation Institution (DRI)." DRI has six main functions:

- Exhibition: Cooperating with people affected by the earthquake, public, volunteer staffs, and other contributors, the experiences and lessons of the earthquake are exhibited in a museum. Especially, the exhibition is expected to make children feel the importance of disaster management and of lives.
- Collection/storage of data and records: To transfer the lessons of the earthquake and the experience of affected people, DRI continuously collects data and records related to the earthquake and disaster management.
- Practical research and cultivation of young specialists: Based on the experiences and lessons of the earthquake, academic knowledge, and accumulated research results, practical research is conducted in order to contribute to policy measures on disaster management for national and local governments, communities, companies, and other groups.
- Field support for disaster response: When large disasters occur, DRI sends staff having practical and comprehensive disaster response knowledge to the disaster-affected area. The staff contributes to damage

reduction and prompt recovery and reconstruction in the area by providing appropriate information to the commanding officer.

- Cultivation of expert staff on disaster management: DRI has several training courses for government officers on disaster management. In the training, the disaster management experiences gained from the earthquake are shared and transferred. In addition, research results and practical knowledge and technology are also provided.
- Intercommunication and network: Various networks among government, researchers, community, companies, and others are made to promote and enhance the coping capacity of society.

As described above, DRI has various roles based on the experiences in and lessons from the Great Hanshin-Awaji Earthquake. DRI can be regarded as an integrated center for disaster management. Maiko High School utilizes the institution for the education of the students. Students listen to affected people to know the situation when the earthquake occurred in 1995. The affected people provide practical knowledge on how we should act before and after a disaster as well as their own experiences. Then students look at the exhibits. In the museum, videos to reenact the collapse of buildings and the intensity of the earthquake are shown. Dioramas are also exhibited to show the city after the earthquake occurred. In addition, other things related to the earthquake are shown in the exhibition area.

Community Safety through School Safety: Case of Nepal

This section introduces the community safety project conducted through schools by the NGO called NSET–Nepal (National Society for Earthquake Technology–Nepal). The project of NSET–Nepal was named School Earthquake Safety Program (SESP). In the case of Nepal, the school buildings are vulnerable against earthquakes, especially the government schools. There is a large difference between government schools and private schools in terms of availability of funds. If an earthquake occurs, private schools can recover by themselves but government schools cannot. Therefore, NSET–Nepal emphasizes government schools to conduct the project.

When NSET–Nepal started SESP in 1997, the emphasis was given to the structural safety of school buildings and enhancing the ability of masons in retrofitting or rebuilding school buildings, because of the following reasons: (1) the school is a central facility for the community, (2) the school is one of the vulnerable buildings, and (3) the school has responsibility in saving future generations (see Fig. 1). The technology that NSET–Nepal teaches is locally oriented and available for all and is not a new technology. Therefore, the masons can build earthquake-safe buildings by themselves after participating in SESP. During the mason training, NSET–Nepal organizes a shake table demonstration in order to give masons confidence on the technology that they have learned (see Fig. 2). In the shake table

Fig. 1. Government School Buildings Rebuilt Through SESP (photo taken by author).

Fig. 2. Shake Table Demonstration for Masons, Policy Makers, Law Makers, Engineers, Technicians, General Public, and School Students to Make Them Understand the Performance of Buildings with Small Improvement in Construction (photo taken by NSET).

demonstration, NSET–Nepal prepares two same-size house models. One is built using retrofitting technology and the other is built using normal local construction method. The model with the normal technology collapses during the demonstration of earthquake shocks, but the model with retrofitting technology will withstand the shocks. Therefore, through this convincing demonstration, the masons can have confidence on the retrofitting technology.

After evolving through several processes, the SESP consists of three parts. The first is retrofitting or rebuilding school building; the second part is training for masons, teachers, students, and community people; and the final part is awareness program for students, teachers, and community. One of the aims of this project is community-based disaster management through these three activities. Masons learn and understand appropriate technology for earthquake safety by constructing or retrofitting school buildings. After the project, they can build earthquake-resistant houses and buildings. As part of SESP, teachers, students, and community members are given information about disaster management. NSET–Nepal also establishes an Earthquake Safety Club as one of the activities in each project school. The club consists of students as its members. This club organizes essay writing and drawing competitions and similar events related to the criticalities of disaster management in each of the schools. By implementing SESP, masons, school teachers, students, and community people are included. These people are the stakeholders of community-based disaster management. Through SESP, they can get common understanding on what is necessary for disaster reduction. Masons, students, teachers, and community people receive technology as well as information, learn the importance of mitigation and preparedness, and take supportive action for disaster reduction.

Co-Learning between the Community and Students: Case of Japan

Saijo City, in the Ehime Prefecture of Japan, was affected by two large typhoons in 2004. In the mountainous area of the city, landslide occurred. Because of the landslide, a lot of sediments and wood that had been put on the surface of the mountain flowed to the river. These sediments and wood were blocked at bridges or at narrow spaces of the river, causing floods in various places in the city (Yoshida, Takeuchi, & Shaw, 2009). The flood resulted not only in human damages including the death of five people but also in damages to residences. In addition, communication between the mountainous area and the plain area was interrupted. Therefore, requests for response or other emergency requests could not be transmitted, and many people were isolated in the affected areas. As the necessary factors in disaster management, the city understands appropriate mountain management, cooperation and mutual understanding between the mountain and plain areas, and disaster education. In 2006, the city started an education program for students in the sixth grade in elementary schools. The age of sixth-grade students ranges from 11 to 12 years. The city named the education program as "Age 12 Education." There are some reasons why the city focuses on 12-year olds. Twelve-year-old students have enough physical power even though it is weaker than that of mature persons. They can make decisions by themselves. And they spend time in the community area.

As part of Age 12 Education, town watching and mountain watching have been conducted in Saijo (see Fig. 3). Participants of town and mountain watching are not only students but also parents, community association members, and local government officials. Involving different stakeholders makes the participants understand the role of each stakeholder

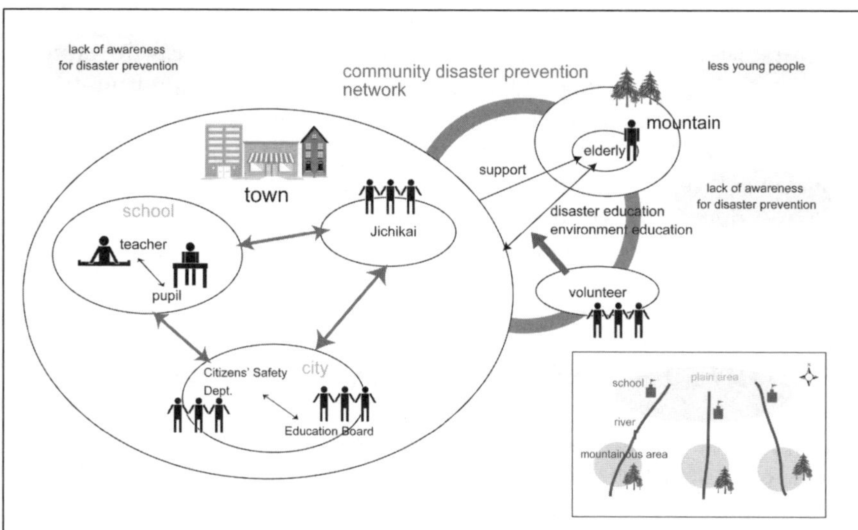

Fig. 3. Concept of Town Watching and Mountain Watching (*Source*: Yoshida et al., 2009).

and their awareness is expected to be enhanced. Therefore, town and mountain watching can be effective activities in community-based disaster management.

Saijo holds the "Saijo Children Disaster Management Summit" under the umbrella of the Age 12 Education program as a concluding event at the end of the fiscal year (Saijo City, 2009). In the summit, students present their learning, experiences, activities, and action plans. Students in the sixth grade from all schools in the city are invited, and the number of the students is more than 1,000. The characteristic of the disaster education in Saijo is the sharing and transfer of disaster experiences. During town watching and mountain watching, affected people can transfer their experiences to the children. Mature people can share their experiences and knowledge of the local context. Involving various kinds of people is an important factor in this regard. To make this kind of education continue, students who do not have disaster experiences can take education from the experiences shared. If the community has people who have disaster experiences, Saijo can achieve direct transfer of disaster experiences from the affected people to the new generations. Another characteristic is that participants understand the roles of other stakeholders through town watching and mountain watching. Community association members can understand the children's views, the local context, and the role of the local government. Local government officials can know the community members' views and share the works or plans of the local government with them. Town watching and mountain watching can be opportunities for communication between local government officials and community members, including the children, toward a community-based disaster management. In the summit mentioned above, many children try to present their own opinions in front of 1,000 participants. It is a great achievement. It can be regarded as the effect of involving various participants in town watching and mountain watching.

EFFORTS IN PROMOTING DISASTER EDUCATION

This section shows several efforts for the promotion of disaster education. In the case of school education, teachers play important roles for the disaster education of the students. In the case of community education, NGOs or community organizations are significant. This section consists of two cases, one focusing on school teachers in Nepal and Indonesia and another featuring an NGO in Pakistan.

Program of Teacher's Training and its Evaluation: Case of Nepal

To promote school disaster education, the roles of teachers are the most significant. A survey on school teachers was conducted in Kathmandu, Nepal. Here are the main findings (Shiwaku, 2007):

- Teachers were implementing disaster education, but it was based on the textbooks used in the curriculum.
- The textbooks included disaster-related topics and environmental topics.
- The disaster-related topics like mechanism of earthquakes or floods were provided and considered as disaster education. But some social studies topics related to disaster management were not provided.

In the training, teachers of science, environment, and social studies were targeted because these three subjects are deeply related to disaster management, compared to other ongoing subjects. The training program had three parts:

- Disaster Management in Nepal
- Concepts and Examples of Disaster Education
- Group Work

One of the important outputs of the exercise was to develop concept map (see Fig. 4). The resulting concept map shows the teacher the extent of the knowledge that the students possess. In the concept map, students can describe words or topics if they understand the relationship with other

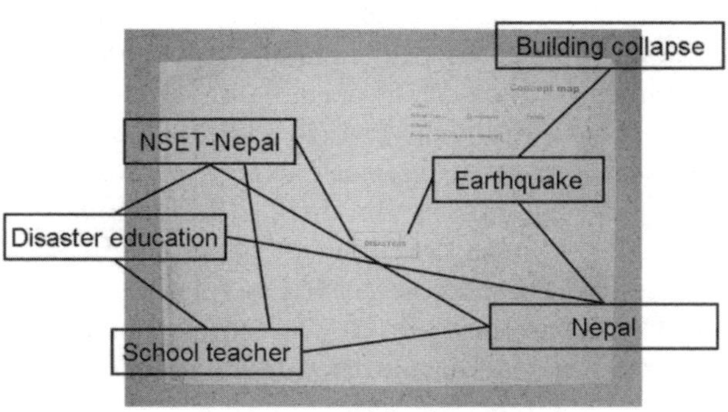

Fig. 4. Using Concept Mapping in the Evaluation of a Teacher's Knowledge on Disaster Management.

words or topics. Therefore, knowledge does not mean that students just know the words. Through the concept map, teachers can measure the amount of knowledge and image or concept of disaster management that students have. For this reason, Maiko High School conducts concept mapping two times in order to evaluate the improvement in the student's knowledge. For the evaluation of the training, concept mapping by the participating teachers was conducted three months after the training. Fig. 4 shows an example of the concept map used in giving instructions to teachers before they start their own concept mapping. In this case, the word "DISASTERS" was provided at the center of the paper. Concept mapping is a useful evaluation tool to know the teachers' understanding and knowledge of disaster management and disaster education. For the person who evaluates training programs, implementing concept mapping also contributes to improvement of the program in the future. In school, conducting concept mapping can make teachers realize their own knowledge and accelerate the discussion among teachers. Given these various uses, concept mapping is effective in several aspects.

Peer Education Among Teachers: Case of Indonesia

SEEDS Asia, which is a nonprofit organization (NPO) in Japan, conducted a project to promote school safety education in Sanden, Bantul District, Yogyakarta, Indonesia, in 2009. The district was an area affected by the Java Earthquake of 2006, so the importance of the disaster education project was validated. The main objective of the project was to build the capacity of school teachers of elementary schools in Bantul District to be able to implement school safety education with confidence to their students at their own school. In this project, the main components were the teachers' training, the implementation by teachers, and the workshop for the promotion of school disaster education (SEEDS Asia, 2009). The teachers' training targeted 43 teachers of three elementary schools. The training had two parts. One was the lecture on disaster management and disaster education. The other was the development of an action plan on disaster education by teachers. The schools needed to integrate disaster education into the school curriculum (see Fig. 5). Therefore, a material showing the link between disaster education and the school curriculum was prepared. Because of the material, the teachers' interest in disaster education was enhanced and teachers could select an education program that could be implemented in their own subjects. The education programs under the

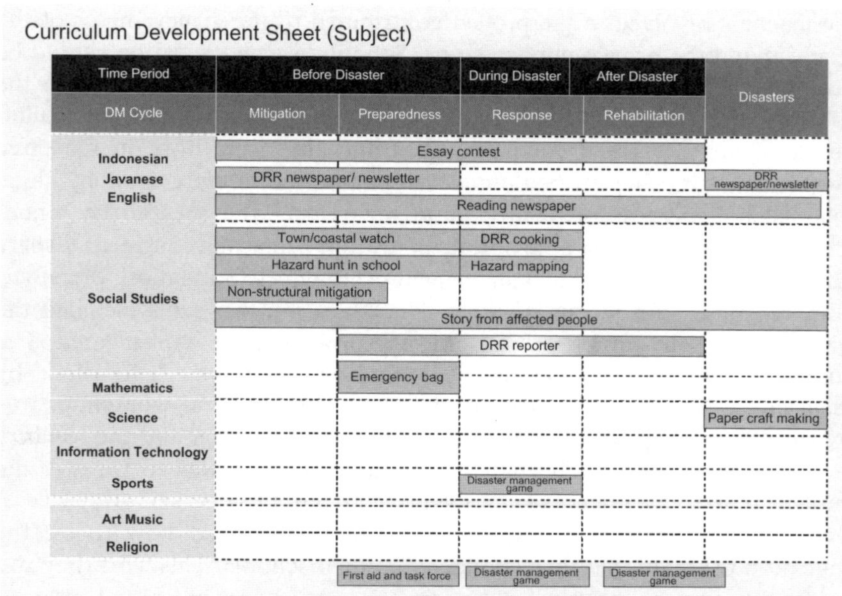

Fig. 5. Sample Curriculum Development Sheet (SEEDS Asia, 2009).

action plan developed by schools were actually implemented. The implementation was done in each school with the support of SEEDS Asia and a local NGO. After the implementation, the teachers from the three schools had a workshop. The purposes of the workshop were to share the teachers' experiences on the implementation of the disaster education in their school and to improve their action plan. Then the implementation of the improved action plan was done without the help of SEEDS Asia and the local NGO. Through this process, the following achievements of the teachers and schools were seen:

- Schools could develop a disaster education curriculum according to their situation or interests.
- Teachers could implement disaster education by themselves.
- Teachers could have confidence on the implementation of disaster education.

Before the project, the teachers could not implement disaster education. The actual implementation by them and the sharing of experiences among the teachers involved in the project contributed to the achievements of the teachers and the project implementers. School disaster education should be promoted in schools in a certain area as well as targeted schools. In the project, a teacher's guideline was developed, and it consisted of curriculum development sheets and education programs that were done in the three project schools. The curriculum development sheets were used by three schools and were subsequently improved. Using the sheets, the school teachers could select the appropriate education programs considering spare time, subjects, and disaster management cycle stages (mitigation, preparedness, response, and reconstruction). The education programs included the experiences in the project schools and the process to be implemented. The guideline allowed the schools to implement disaster education by themselves. To promote school disaster education, a workshop was conducted, inviting school teachers in Kecamatan Sanden and the teachers in the project schools. The purpose of the workshop was to transfer the implementation experiences from the project school teachers to other school teachers. The teachers of the project schools explained how to do the education program that they conducted in their schools. This workshop did not involve knowledge transfer from specialist to learner but from learner to learner. The workshop had the following effectiveness:

- The teachers of the project schools had confidence that they could teach how to conduct disaster education.
- Teachers of other schools thought they would also be able to conduct disaster education because teachers at the same level as them had conducted disaster education.

Generally, before the workshop, many of the teachers felt some difficulty in conducting disaster education because of their lack of knowledge on disaster management. The learning among teachers in the workshop solved this problem.

Capacity Building of NGO Leader as Community Organizer

After the Pakistan Earthquake of 2005, many projects were conducted in the affected areas in the field of human development and disaster management (Shiwaku, 2009). The main purposes of the project (conducted by the first author) were to provide training programs to local people including the local government and to establish a knowledge center. The roles of the knowledge center were to provide information on disaster management and to collect and transfer village information on livelihood. This means that the knowledge center was the community center for disaster management. In the project, three Village Knowledge Centers (VKC) and District Knowledge Center (DKC) were established (UNISDR, ADRRN, & Kyoto University, 2007). VKC is managed by a VKC committee and a local NGO. DKC is managed by the local government and a project assistant. In the project, three training programs (community-based disaster management, safer construction, and disaster education) were provided and involved the local government, local NGOs, and local people. The implementation process of the project included five main activities (UNISDR et al., 2007):

- Meeting with the local government
- Needs assessment survey/focus group discussion
- Site selection of VKC
- Implementation of training programs
- Establishment of DKC and VKC

The main achievements were the following:

- Awareness raising of local government, local NGOs, and local people on disaster management
- Establishment of VKC and DKC
- Communication among stakeholders

During the project, the author employed a college student as the project assistant. At the end of the project, the assistant established his own NGO called the Sustainable Development and Response Foundation (STAR) and became its president. STAR started their own programs including the management of DKC and the distribution of brochures developed during the project. The activities of STAR were divided into two types. One type consisted of activities based on the project in which the president was involved as a project assistant. The other type consisted of activities that STAR developed on its own or that the local government requested. The roles of the NGO could be regarded as that of trainer and coordinator in the community. UNCRD (2004) considered policy makers, national disaster managers, local disaster managers, trainers, and community workers as main actors in community-based disaster management. The activities of STAR allowed it to play the role of trainer and community worker. The work areas were the remote area and mountainous area. Since the local government did not have high awareness of disaster management, the activities of STAR were very important to the communities. As mentioned above, the president of STAR was just a college student. It means that he did not have specialized knowledge on disaster management and he did not know how to conduct awareness-raising programs before being involved in the project. But STAR is contributing to the local community now. The positive changes he underwent as he got involved in more activities can be regarded as the impact of a good practice in capacity building for disaster reduction. Based on his comments, a capacity-building process can be proposed as follows:

1. Motivation toward disaster reduction
2. Confidence building
3. Making connections or networking

Next, the effective factors that can achieve each stage mentioned above are examined and discussed.

1. Motivation toward disaster management: Through the activities mentioned, the assistant could get knowledge on the project and disaster management. At the beginning of the project, his main task was as interpreter between the project manager and local government officials and villagers. He was required to get concrete knowledge needed not only to be able to translate but also to explain. Discussion was always done between the project manager and the assistant. Discussion was effective to transfer knowledge on disaster management. Discussion was a good opportunity for the assistant to consider what was useful or advantageous for the local people. Shiwaku and Shaw (2008) emphasized that discussion is a process in disaster education. Capacity building can be regarded as a kind of education. The project assistant realized the importance of disaster management because of his close discussion with the project manager.
2. Confidence building: From the middle of the project, the assistant's tasks were increased and his responsibility was also increased. He had opportunities to negotiate with the local government or local villagers. These opportunities required him to have more concrete knowledge and negotiation skills. Through the negotiations, he could make a strong connection with local government, local NGOs, and local people. In addition, his work made the project a success. Through his activities, he was able to build self-confidence. It is important not only to transfer knowledge but also to let a person have responsibility in the field.
3. Making connections or networking: Through the project, the assistant negotiated and discussed with different persons, including someone from UN ISDR. If he was not involved in the project, he could not meet such

people. It was a great opportunity for him to meet them, and the meetings motivated him to be involved in disaster management. In addition, he could make connections with various people. He could conduct many activities, utilizing his connection with stakeholders. To continue NGO activities, help or cooperation from stakeholders is necessary. Additionally, making good relationship or meeting with different persons difficult to meet in daily life was a great opportunity for the project assistant. Such opportunities pleased him and motivated him more.

Onda (2001) mentioned that capacity building includes conscientization and self-empowerment. According to his description, self-empowerment is the basis of self-reliance, self-help, and self-determination. The three steps proposed as capacity-building process in this chapter have similarities to conscientization and self-empowerment. Motivation toward disaster management is conscientization. Confidence building is self-reliance. Through making connection, the NGO could conduct their activities. This process is self-help and self-determination. Therefore, the capacity-building process for STAR can be regarded as a good practice.

SHARING DISASTER EDUCATION MATERIALS ON THE INTERNET

Before the advent of the Internet, acquiring and sharing disaster education materials was slow, limited in scope, and expensive. Contacting the relevant people who produced the materials to ask for clarifications or obtain more details was also quite difficult, especially if the materials originated from another part of the world. But today, thanks to advances in information technology, many disaster education materials can be conveniently shared and posted on websites. In this chapter, among the numerous resources on the Internet, we are presenting four sample websites where sharing of disaster education materials can and do take place. The first one is a UNISDR-initiated project intended to reach out to a wide range of audience, disaster risk reduction (DRR) professionals and nonspecialists alike. The second one is an academe-GO-NGO collaboration specifically targeting DRR practitioners. The third website is an initiative especially aimed at primary and secondary school students and teachers for sharing community safety maps. Lastly, the fourth one is a social network dedicated to disaster prevention education.

PreventionWeb

In November 2007, in support of the Hyogo Framework for Action (HFA), the UNISDR Secretariat launched PreventionWeb (http://www.preventionweb.net/english/), a website for increasing knowledge sharing on DRR issues for both DRR practitioners and the general public at all levels, from local to global (UNISDR, 2007a). PreventionWeb caters to everyone interested in reducing disaster risk: government officials, disaster managers, development and humanitarian workers, media professionals, teachers and students, etc. PreventionWeb relies on contributions from the DRR community, including UN, international, nongovernmental, academic, and civil society partners, and includes DRR news, country reports, publications, good practices, fact sheets, directories, networks and communities, policy documents, training events, jobs, terminologies, country profiles, audio and video contents, and more (UNISDR, 2007b). The site is updated daily. PreventionWeb has a special section for the "Education and School Safety" theme and also has a segment containing education materials for DRR and information on academic programs in DRR, organized under "Professional Resources." When it was still being designed, it was suggested that PreventionWeb could be considered a success if it would be considered *the* reliable place for information on disaster risk reduction. However, it is admitted that it is really the use of information that can be proved to reduce disaster risk at the local level that is the true test of PreventionWeb.

Disaster Reduction Hyperbase

An international initiative called the "Disaster Reduction Hyperbase – Asian Application (DRH-Asia)" was initiated based on a proposal by the Japanese government at the UN World Conference on Disaster Reduction (UN-WCDR) in January 2005, to promote the development of a disaster reduction portfolio that could be an effective information platform for disaster risk reduction (Kameda, 2011). DRH (http://drh.edm.bosai.go.jp/) is an interactive vehicle for collecting and sharing appropriate disaster reduction technology and knowledge designed for the potential use by policy makers, community leaders, practitioners, and researchers who wish to access appropriate technical know-how's that can help them establish practical disaster management plans. Shiwaku (2011) proposed other utilization of disaster reduction technology and knowledge registered in

DRH-Asia. The technology and knowledge can be used for education and training. Through the implementation of the technology and knowledge by NGO, the implementation process becomes on-the-job training for NGO. For educational use, the technology and knowledge provide practical information. Therefore, university students can combine such information and theoretical knowledge through learning from the website.

Natural Disaster Youth Summit

The Natural Disaster Youth Summit (NDYS) led by JEARN (Japan Education and Resource Network) started as a gathering called the "Natural Disaster Youth Summit 2005 in Hyogo," which invited children in other countries, in commemoration of the 10th anniversary of the Great Hanshin Awaji Earthquake. The summit was intended to promote awareness of the devastation caused by disasters and to impress the importance of saving lives in the minds of young children, based on their experiences gained from similar natural disasters in recent years. These lessons should be passed from generation to generation. The purpose of NDYS is promotion of disaster education through international exchange. The main objective of international exchange learning would be the understanding of different cultures, to raise awareness that human beings are in fact global citizens. NDYS also realizes that globalization is already happening and many problems cannot be solved domestically. This is why international exchange learning is important. The education of NDYS requests students to make disaster safety map as one-year education program. Students learn about past disasters and present states of disaster management in their community, learn to imagine problems when disasters occur, and learn to think of measures for mitigation and preparedness. These learning processes are facilitated by students' discussion using a map of their school and surrounding areas. Specific measures for mitigation and preparedness are written by students on the map and the map is finalized as a Community Disaster Safety Map. The map of each school is collected and connected to become a disaster safety map of the world, which can be called the Global Disaster Safety Map. In the process of the education, students use the forum provided by NDYS website (http://ndys.jearn.jp/eng.html). Students can share and exchange information about their learning process and achievements. Using the forum, students can learn disaster management among students in other countries. The school-level maps prepared by children can then be fed into higher-level mapping. Numerous maps made by students from different countries are already on the NDYS website and the collection is expanding every year. Maps can record and display risk information clearly and conveniently. Maps are a good medium for risk communication, and as NDYS slogan says, "Communication saves lives!"

Coalition for Global School Safety and Disaster Prevention Education

The Coalition for Global School Safety and Disaster Prevention Education (COGSS & DPE) is an international network whose mission is "to identify gaps and priorities and to support the development of knowledge-sharing strategies and political will to ensure that every school is a safe school, and that every child and community has access to high quality, audience-targeted disaster prevention education knowledge, experience, and expertise to build a culture of safety" (PreventionWeb, no date). COGSS & DPE (http://cogssdpe.ning.com/) has four core working groups: Disaster-Resistant School Infrastructure, Disaster Prevention Education in Schools through Formal Curricula, Disaster Prevention Education: Extra-Curricular and Community-Based, and School-Based Disaster Management. COGSS is devoted to complement, not substitute or duplicate, any ongoing institutionalized disaster risk reduction efforts (Bender et al., 2007). Membership to COGSS & DPE requires the approval of the administrator. As of December 2010, the network had more than 400 members. As its name implies, the COGSS & DPE website focuses on school safety and disaster prevention education. There are downloadable e-newsletters, a directory of members, photos and videos, blog posts, a forum section, events announcements, and a chat feature, and recently a School DRR Global Activities Map had been added. Links to additional working groups and outreach are also available: Teachers Network, Children & Youth Participation, Disability & Disaster Prevention, Disaster Resistant Universities, Capacity-Building: Higher Education and Young Researchers, Edu4DRR Teachers' Network, etc. The availability of all these websites specializing and focusing on the important DRR issue, that is, disaster education, makes the lives of teachers and educators easier and more convenient, allowing them to focus their energies on the effective delivery of disaster education lessons to their students in and out of the classroom. As a consequence, given these enormous amount of useful and helpful resources made accessible on the Internet, disaster education ceases to look like an overwhelming task.

CONCLUSION

This chapter discusses some approaches that can make disaster education more effective. This chapter shows several cases of disaster education using school-community linkage. These cases can be classified into the following:

- Students learn in the community (through town and mountain watching)
- Students and community members learn together (through interview and sharing of disaster experiences)
- Community members, teachers, and students learn under the same purpose

Efforts for the promotion of disaster education can be classified into the following:

- Providing training to teachers
- Co-learning among teachers (through a workshop where teachers previously provided with disaster education training shared what they learned with teachers who were not trained)
- On-the-job training (for example, the masons in Nepal and an NGO president in Pakistan)

With regard to the Internet, each disaster education website has several roles. The following are the proposed roles of websites in order to make disaster education more effective:

- Sharing information in order to learn from others
- Providing information to promote implementation of disaster risk reduction and disaster education activities
- Collecting information useful to information users

In the examples presented in this chapter, there is a characteristic point. The examples show that a specialist in disaster education is not always necessary in order to be able to conduct and promote disaster education. Disaster education can be conducted and promoted using innovative approaches.

One of the issues in disaster education that should be considered is sustainability. Sustainability is regarded to have two aspects. One is quality of disaster education and the other is the availability of human resources. School–community linkage is expected to contribute to both. Through learning with community members, school students can understand how they should act in the community. Then they can facilitate disaster education, which involves the school and community after they mature.

Community people become aware of disaster management through such education program. In addition, they can share and transfer their disaster experiences to students. School-community linkage can assist in promoting community-based disaster management and in contributing to the sustainability of providing quality disaster education and in ensuring the availability of trained human resources working in disaster education. School teachers and NGOs should be included in efforts promoting disaster education. These efforts can contribute to both aspects of sustainability mentioned earlier. Teachers and NGOs can organize disaster education program by themselves, considering the local context. It is possible that after some years have passed, the local context may change. For example, 10 years after a disaster, community members with disaster experiences may still be around. But the community would then have young students without such experiences. The education provider should consider the difference between those with and without disaster experience. Therefore, it is necessary to train persons who can arrange appropriate disaster education programs as well as provide knowledge on disaster management.

To support the two aspects of sustainability, the Internet is useful. The Internet can facilitate the sharing of good practices around the world. Trained teachers or community leaders can find various kinds of helpful information on the Internet, and hopefully they will actually utilize them. However, in some areas of the world, it is not always possible to use the Internet. In such areas, it might still be necessary to provide printed documents that contain good practices. In addition, it should be noted that not only using information but also collecting information on good practices is important in order to provide diverse kinds of information on the Internet that can be applied in different situations and contexts. To provide disaster education that is more effective and sustainable, appropriate approaches should be utilized. Fortunately, there are a variety of options available.

ACKNOWLEDGMENTS

This chapter compiles the results of several researches. Mr. Seiji Suwa, a teacher of Maiko High School, provided much information about the Environment and Disaster Mitigation Course. NSET–Nepal also provided information and materials on their project. Saijo City, Japan, cooperated with the survey on town and mountain watching. SEEDS Asia and STAR Foundation provided their project sites for the survey in Pakistan. In addition, school students and teachers, and local community members, warmly cooperated with the researchers. They are all gratefully acknowledged. We would also like to extend our gratitude to the Ministry of Education, Culture, Sports, Science, and Technology (MEXT) of Government of Japan scholarship and the Environmental Management Leaders (EML) Program for funding second author's internship programs.

REFERENCES

Bender, S., Wisner, B., Petal, M., Kumar, H., Pomerai, G., Gupta, M., & Izadkhah, Y. (2007). Coalition for global school safety (COGSS) profile. Available at http://www.interragate.info/cogss. Retrieved on December 22, 2010.

DesInventar (no date). Available at http://www.desinventar.net/DesInventar/index.jsp. Retrieved on January 23, 2011.

Kameda H. (2011). Disaster Reduction Hyperbase (DRH) – conceptual development and production. In: H. Kameda and K. Shiwaku (Eds), *Disaster reduction technology information: DRH initiative toward implementation of product, process, and wisdom, Asian Journal of Environment and Disaster Management*, *3*(1), 5–19.

Onda M. (2001). *Development sociology: Theory and implementation.* Tokyo, Japan: Minerva Publishing. [in Japanese]

Petal, M. (2009). Education in disaster risk reduction. In: R. Shaw & R. Krishnamurthy (Eds), *Disaster management: Global challenges and local solutions* (pp. 321–337). Hyderabad, India: Universities Press.

PreventionWeb. (No date). Coalition for Global School Safety and Disaster Prevention Education (Organization Profile). Available at http://www.preventionweb.net/english/professional/contacts/profile.php?id=3748

Saijo City. (2009). Saijo Children Disaster Management Summit. Available at http://www.city.saijo.ehime.jp/khome/gakkokyoiku/12saikyoiku/index.html. Retrieved on December 22, 2010. [in Japanese]

SEEDS Asia. (2009). School Safety Education in Bantul District. Available at http://www.seedsasia.org/eng/indonesia_school_educationJPFen.pdf. Retrieved on December 22, 2010.

Shaw, R., & Takeuchi, Y. (2008). Environment and disaster management – Role of people and community. *Series of Kyoto University's lecture – Approach to environment of globe, Workshop on global environment of Kyoto University* (pp. 45–57). Tokyo, Japan: Maruzen Publishing.

Shiwaku K. (2007). Towards innovation in school disaster education: Case research in Kathmandu, Nepal. PhD Thesis, Kyoto University.

Shiwaku K. (2009). Reconstruction Project in the Affected Area of Pakistan Earthquake. Summaries of Technical Papers of Annual Meeting, Architectural Institute of Japan, F-1, 557–558.

Shiwaku K. (2011). Innovative usage of disaster reduction technology information. In: H. Kameda & K. Shiwaku (Eds), *Disaster reduction technology information: DRH initiative toward implementation of Product, Process, and Wisdom, Asian Journal of Environment and Disaster Management, 3*(1), 107–118.

Shiwaku, K., & Shaw, R. (2008). Proactive co-learning: A new paradigm in disaster education. *Disaster Prevention and Management, An International Journal, 17*(2), 183–198.

Shiwaku, K., Shaw, R., Kandel Chandra, R., Shrestha, N. S., & Dixit, M. A. (2006). Promotion of disaster education in Nepal: The role of teachers as change agents. *International Journal of mass emergency, 24*(3), 403–420.

UNCRD. (2004). A USER'S Guide: Sustainable Community Based Disaster Management (CBDM) Practices in Asia. UNCRD publication.

UNISDR. (2007a). PreventionWeb: A new tool to increase knowledge on disaster risk Reduction. Available at http://www.unisdr.org/eng/media-room/press-release/2007/pr-2007-17-PreventionWeb.pdf. Retrieved on December 22, 2010.

UNISDR. (2007b). PreventionWeb Project. Available at http://www.preventionweb.net/english/documents/about/PW_project_description_GP.pdf. Retrieved on December 22, 2010.

UNISDR, ADRRN, and Kyoto University. (2007). Final Report on Training and Capacity Enhancement of Local Governments in the Earthquake Affected Area of Pakistan, 26–32.

Yoshida, Y., Takeuchi, Y., & Shaw, R. (2009). Town watching as the useful tool of urban risk reduction: Case of Saijo, Japan. In: R. Shaw, H. Srinivas & A. Sharma (Eds), *Urban risk: An Asian perspective* (pp. 189–206). Bingley, UK: Emerald Publication.

CHAPTER 7

IMPLEMENTATION TOOLS FOR DISASTER EDUCATION

Farah Mulyasari, Yukiko Takeuchi and Rajib Shaw

INTRODUCTION

Following the adoption of the Hyogo Framework for Action, various disaster educational materials (UN/ISDR, 2006) that are described as "tools," taken in various forms such as in printed materials (booklets, leaflets, textbooks, handbooks/guidebooks, and posters) and nonprinted materials (activities, games, and practices) were developed. These tools have an important function in communicating the disaster education to the public via formal, non-formal, and informal education, which may take place at school, at home, and/or within the community. In addition, media may also serve as a communication tool. Talero (2004) proposed that the modern communications nowadays have provided information for the growing public demand for related information, which can be used as educational aids to reduce the gap between scientific knowledge and civic awareness.

Shaw and Takeuchi (2009a) proposed that disaster education should go beyond textbooks and involve experiential learning. Textbooks may provide the knowledge; however, it is more important to transform the knowledge into practice, and this would involve practical training, capacity building, and experiential learning. Town watching/regional and/or community watching are considered as key tools for this purpose. In the 2007 August

Disaster Education
Community, Environment and Disaster Risk Management, Volume 7, 137–151

ISSN: 2040-7262/doi:10.1108/S2040-7262(2011)0000007013

mid-term report from MEXT, Japan, it was stated that disaster education should achieve the following four abilities: (1) the ability to take action in disaster preparedness and mitigation by knowing local disasters and community characteristics of each residential area and acquiring disaster prevention technologies, (2) the ability to protect oneself from natural disasters and to cope with postdisaster life if one suffers from disasters, (3) the ability to rehabilitate disaster-affected areas and to reconstruct secure and safe society, and (4) the ability to voluntarily help other people and areas to be safe. Education for disaster aims to accelerate the progress of societies toward disaster resilience and, at the same time, increase awareness and develop proper knowledge and skills among the individuals. To achieve this, appropriate and effective tools are necessary. These tools will enable the individuals to act -- individually and collectively – to cope with disasters.

This chapter will describe the existing tools, the users and examples of the implementation of the tools as good practices in disaster education, as well as discussion on how to deal with and enhance the tools for better disaster education process in the future. Since the turn of the millennium, the impact of communication and information sharing on disaster has become increasingly important. Modern communication and information sharing platform tool, via the Internet, has proven to be a critical tool in disseminating information (i.e., disaster education) to the public.

CURRENT DISASTER EDUCATION TOOLS

The disaster education tools can be considered based on the following: (i) mode of education (i.e., formal, non-formal, and informal education) and (ii) type (i.e., printed, nonprinted, etc.). The integration of both formal and informal education through schools is strongly encouraged to ensure that disaster education reaches every home and community and that learning is sustained into future generations (Petal & Izadkhah, 2008). Formal education (i.e., disaster education in curricula) supplemented with non-formal (e.g., museum visits) and informal (e.g., experiences from family members) education is recommended. Tables 1 and 2 show examples of the tools used in formal and informal education, respectively.

Table 3 provides an overview of the various tools with special emphasis on the users of the disaster education tools.

Table 1. Tools Used in Formal Education.

Approach	Details of the approach
Curriculum integration	• Makes use of specially developed units, modules, or chapters concentrating on disaster risk reduction. • Designed to fit into specific course curricula, grade level, and duration. • Teachers training is necessary to development competence and efficacy
Extracurricular integration	• Campaigns with local governments, which could also help increase public awareness.
Curriculum infusion	• A more comprehensive approach that distributed disaster risk reduction content throughout the curriculum, using lessons, readings, and activities, enriching the existing curriculum. • Require high-level policy guidance, resources, collaboration between curriculum specialists and disaster risk reduction experts in developing and evaluating the curriculum. • Full curriculum adoption may take 5–10 years
Stand-alone course	• Specialized course curricula focused on disaster risk reduction
Curriculum resource materials	• Strategy that has been used in California and throughout the United States of America (American Red Cross, 2008; Team Safe-T, 2008) • Teachers are permitted the flexibility to select materials, a wide access of which is facilitated through Internet delivery and where a large numbers of volunteers make themselves available to support lessons and projects in schools.

Source: Petal (2008).

USERS OF DISASTER EDUCATION TOOLS

Before identifying the users of disaster education, the stakeholders within community at large can be divided into two groups: those coming from inside the community and those coming from outside the community. In addition, the stakeholders are also classified either as expert in local issues or as expert in disaster prevention and environmental management. The four stakeholders identified are the local residents (A), including the school children, teachers, women, other family members, and communities, who are the community end users, and the community leaders (B) who are the decision makers in

Table 2. Tools Used in Informal Education.

Approach	Details of the approach
Dissemination of written materials	• Posters and signage • Educational tools such as toys and games, documentary and short videos, storybooks, comic books, puzzles, and computer games, etc.
Cultural and performing arts (Bhattia, 2006, in Petal and Izadkhah, 2008)	• Tools such as through music, song, poetry, dance, puppetry, magic, street theater, until improvisation, pantomime, or artworks • May take place at assemblies and special events
After-school "safety clubs," scouting badges, and project activities	• Provide an opportunity to develop awareness materials and displays, plan games, and engage in performances and art projects to communicate with others. • The shake-table demonstrations, which are small-scale models, are also powerful hands-on tools.
Community-service-oriented clubs	• Aims to bring students into contact with local community and local government • Develop students' analytic and problem-solving skills, as they research and identify hazards, tap into indigenous knowledge, oral history, public information, and scientific research and expertise to assess risks and identify solutions.
Competitions, awards and commendations	• Voluntary drawing and writing competitions to engage children as the target of disaster education • "Disaster Risk Reduction Knowledge Tournaments" on radio or television broadcast • Sports day activities, e.g., drills and demonstrations, as well as for competitive and knowledge games like water bucket brigade competition, fire extinguisher target practice, and injury transport relay, which could introduce cooperative response skills
Parents and local community involvement	• Parent–teacher association or school welfare committee meetings, wider community fairs, and "open house" are all tools that withhold important opportunities for informal education. • Exhibitions and displays of student-created risk and capacity maps, models, artwork, and essays are adding personal values and make those tools even more powerful

Source: Petal and Izadkhah (2008).

Table 3. Overview of Disaster Risk Reduction Education Tools.

By Categories		Tools
Learning Process	through Lecture	• Printed materials (textbooks, comics, booklets, leaflets, handbooks, posters, working books)
	through Experience	• Town watching • Interviews • Visit to museums • Disaster drills
	through Experience	• Workshop tools (actions oriented planning)
Approaches in education formats	In formal education	• Curriculum integration • Extra-curricular integration • Curriculum infusion • Broad range of course • Stand-alone courses • Curriculum resource materials
	In informal education	• Dissemination of written materials • Cultural and performing arts • After-school "safety clubs," scouting badges, and project activities • Projects that bring students into contact with local community, local government, and community-service-oriented clubs • Competitions, awards, and commendations • Parents and local community involvement • Participation of community partners • Disaster drills

the community, which is the governments (C), who are the implementer, manager, and resource support provider, and the researchers/NGOs who are the researchers, implementers, education support providers, and plan support providers (Fig. 1). Usually stakeholder (C) does not have direct communication with stakeholders (A), (B), and (D). In some cases, extension workers (like agriculture extension, forest extension, and health extension as well as education extension services) communicate directly to the local communities. In general, the government or stakeholder (C) communicates through the mass media or these extension workers. However, the government is an

Fig. 1. Stakeholders of Disaster Risk Management.

important decision maker; therefore, the other stakeholders have to pay attention and understand information coming from the government.

Additionally, to enhance the understanding of and interaction in community-based education, risk communication is of prime importance. Risk communication is one of the support systems of risk management. The basic risk communication framework developed by Kikkawa (1999, in Shaw & Takeuchi, 2009a) elaborates on one side the information sender and on the other side the information receiver. Both can be also described as the users of the disaster education (Fig. 2). In the framework, the information sender is the government, university, or research institution. They possess much specialized information about disaster education. The information receivers are the communities and individuals that posses many information of the local context.

Risk communication consists of understanding and making action plans. For risk communication to be successful, three items are needed: holistic learning, facilitation, and trust. Holistic learning is the most important in term of risk communication. Therefore, the communication does not make ends meet when there is a gap between the information sender and the receiver; each stakeholder cannot share purpose, understanding, and information. That is why disaster education tools and some risk communication support tools are necessary for the users of disaster education.

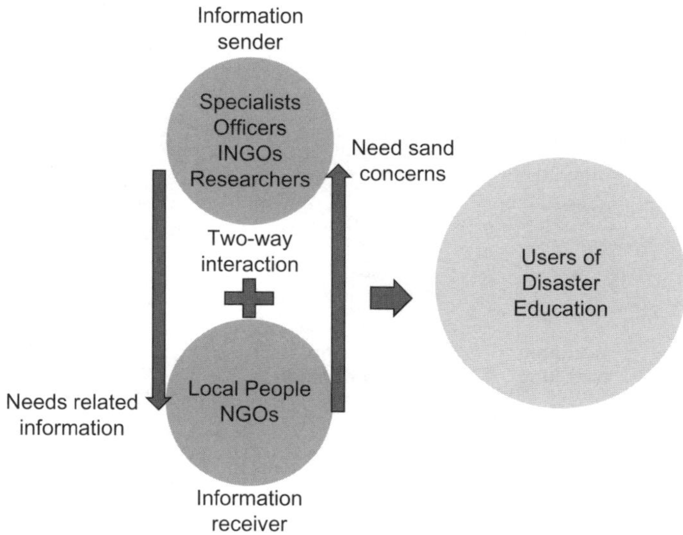

Fig. 2. Users of Disaster Education in Framework of Risk Communication.

GOOD PRACTICES OF DISASTER EDUCATION TOOLS

This section describes selected examples of the good practices that are currently used in disaster education.

Town/Regional Watching

Fig. 3 shows the regional watching concept in the understanding of river basin. Three different concepts of community watching are introduced: (1) mountain watching, to understand different elements in mountains that are linked to disasters and environmental issues; (2) town watching, to understand different elements of the plain land and town and/or villages, which are linked to disasters and environmental issues; and (3) coastal watching, to understand different elements of the coastal areas that are linked to disasters and environmental issues. The three elements provide an overall perspective of regional watching or river basin understanding from the mountains to coastal areas.

Town watching is originally used as a tool for urban planning, but nowadays it can also be used for disaster education. Town watching is

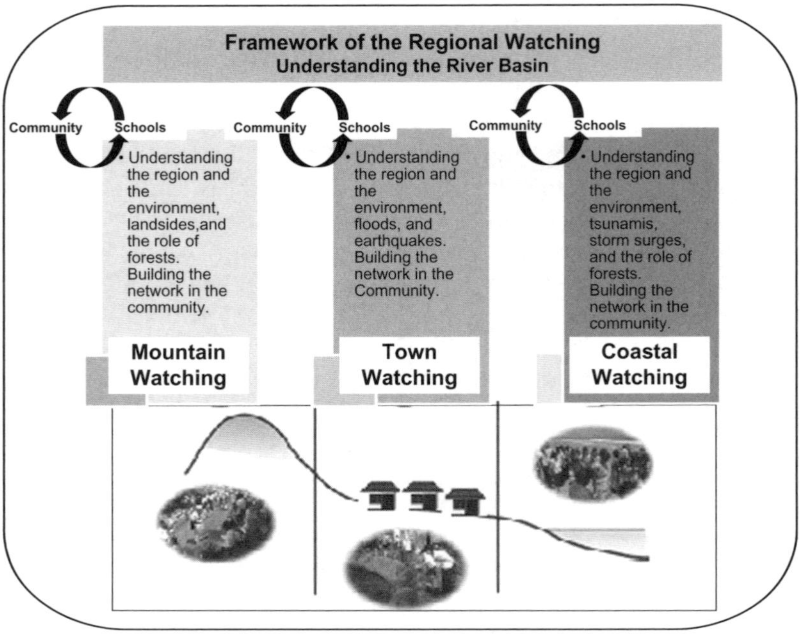

Fig. 3. Understanding the River Basin in the Regional/Community Watching Context.

supposed to be initiated by students in elementary and junior high school and facilitated by teachers, parents, municipal employees, community workers, and volunteers (Shaw & Takeuchi, 2009b). Through these activities, establishing school-based system of mutual help becomes another objective of town watching. The objectives of town watching are as follows: (1) knowing the current situation of the area, (2) increasing children's and adults' awareness of disaster prevention, (3) cultivating children's comprehensive skills and abilities of information collection, thinking, judgment, expression, and communication, (4) telling the experience of disaster victims, (5) pointing out regional problems and suggesting solutions, (6) establishing cooperation system whenever disaster occurs, (7) local residents preparing for disasters with raised awareness of disaster prevention, and (8) becoming a trigger for children to be important leaders in disaster prevention in the region.

Town watching is already mentioned in Chapter 6 as one of the approaches for disaster education. In the case study conducted at Saijo City,

Ehime Prefecture, Japan, that was described, the activity focused on co-learning between the community and students in the mountainous area. The area is affected by large typhoons and landslides, as the collateral hazards. In year 2004, due to uninterrupted heavy rainfall, it triggered severe sedimentation, resulting in serious landslides. Bridges collapsed and sedimentation resulted in flooding as well. During the event, communication between the mountain and plain areas was interrupted; thus request on emergency responses could not be conveyed and many people were isolated in those affected areas. Due to this experience, in 2006, Saijo City initiated the education program, named "Age 12 education," for students of the sixth grade in the elementary schools at the age of 11 and 12 years. And town watching was part of the "Age 12 education." The aim was to understand the role of each involved stakeholder and participants including students, parents, community association, and local government. This is an interactive activity, in which the students with other participants collectively observed and identified the high-risk areas in the city. It provided an opportunity for the risk communication and disaster education between local government, community, and students.

Besides town watching as an approach, it can also be as a tool for disaster education. It is necessary in the future for disaster education to focus on the process of restructuring "communities of practice," not only on the transfer of knowledge and skill between individuals but also as an important goal of education to establish participatory learning in communities, in which both the educator and the learner can "participate" together (Yoshida, Takeuchi, & Shaw, 2009). Such participatory learning is shown in town watching. Town watching can also be used as a tool to measure the impact of disaster education (before and after the activity). A questionnaire survey was conducted before and after the implementation of the activity and results showed changes in the consciousness in (a) hazard, (b) physical impact, (c) damages in the infrastructure, and (d) impacts on people (Fig. 4).

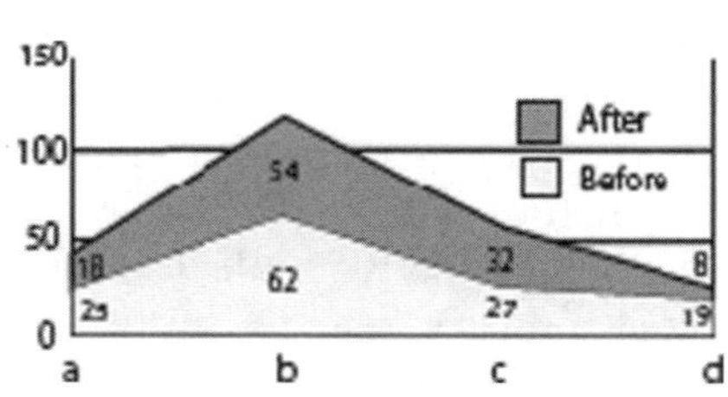

Fig. 4. Example of the Questionnaire Result about the 2004 Typhoon disaster.

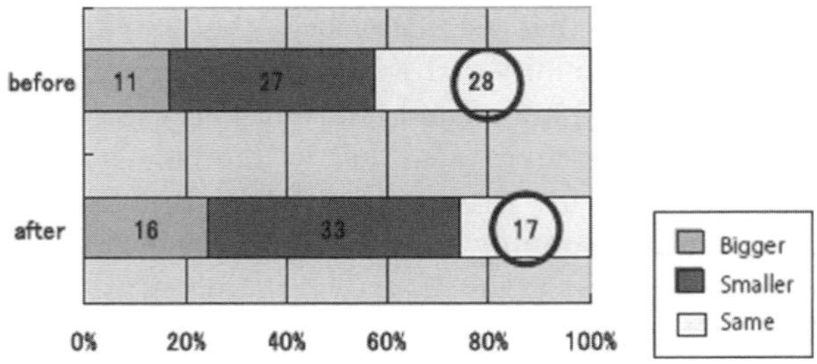

Fig. 5. Example of the Questionnaire Results about the Damage Perception.

Changes were also observed in the damage perception among the individuals (Fig. 5). For example, in the case of one elementary school, the number of "same" decreased from 28 to 17, after implementation of the town watching activity (Fig. 5), suggesting that risk perception among students is enhanced.

GEOMOBIL: THE MOVING VEHICLE OF DISASTER RISK REDUCTION EDUCATION MATERIALS/LIBRARY

The Management of Georisks NAD project (ManGeoNAD, 2008) is a technical cooperation project between *BGR* (Federal Institute for Geoscience and Natural Resources of Germany), the Geological Agency Bandung–-Indonesia, and the Regional Department for Mining and Energy in the Province of Nanggroe Aceh Darussalam (NAD), Indonesia. The ultimate aim of the project is to reduce the impact of future natural hazards on the population following the tsunami and earthquake of December 26, 2004. ManGeoNAD provided recommendations based on geological information (e.g., building ground stability, earthquake hazard) to planners and decision makers to support effective risk mitigation from the beginning of the planning process. In terms of raising awareness, the project developed a strategy to help to build up a culture of preparedness in a simple, straightforward, and cost-effective way. An old eye-catching orange-black Volks-Wagen (VW) van built in the 1950s, the so-called GEOmobil, is used by the project, to visit primary schools in the province NAD. The GEOmobil team comprises experts from the Department for Mining and Energy, provides in-class programs for children on disaster risk reduction,

Fig. 6. The Activities of GEOmobil (MANGEONAD, 2008).

and provides information for the teachers about natural hazard. The VW van serves as a mobile library providing books on natural hazards and is equipped with age-based experiments related to natural hazards. A visit of the GEOmobil team generally consists of a half-day program for the school children from grades 4 to 6. Easy-to-understand and child-centered scientific background on natural hazards is transferred to the children and the teachers with the help of a number of games, songs, experiments, and movies on these occasions. Evacuation drill is often conducted as a form of earthquake preparedness education.

Children are the focus of the GEOmobil activities (MANGEONAD, 2008) due to several reasons: First, children are one of the most vulnerable groups. Second, they are also the future society and the decision makers of tomorrow. Third, children have a strong influence on their parents, disseminate disaster knowledge to their communities, and are, in general, very enthusiastic and receptive in learning about their safety. The experiences with the GEOmobil enhanced children's knowledge of the natural hazards, and helped raise their preparedness and response awareness so as to better cope with the related risks in the future (Fig. 6).

In 2007 the GEOmobil team visited 11 primary schools in the Banda Aceh area in Indonesia and supported the "Disaster Risk Reduction Day" initiated by UNDP in October as well as several exhibitions and fares at the regional and national levels. A total of approximately 1,300 children were directly reached by those activities. Apart from the school visits, a drawing competition was held successfully with 30 children from grade 1 to 6, and a "georisk" calendar for the year 2008 was printed. The calendar contained information on natural hazards, their causes, and things to do before they strike, and was distributed in Banda Aceh. At the beginning of year 2008, the GEOmobil team also finished the production of a movie titled "Disasters and Us: Understanding Earthquakes," which consists of different sequences related to earthquakes, e.g., a picture story drawn by children from Aceh, scientific explanations by the Meteorological Agency

NAD (BMG NAD), earthquake experiments, facts and fictions, and preparedness actions, etc. The movie will be distributed to schools that are not reached by the GEOmobil team. This was a pioneer attempt to reach teachers and children in poor accessibility areas. In addition, trainings for teachers in at least 60 schools in NAD are planned, and a closer cooperation with the Department for Education is envisaged. A second van began functioning in May 2008. The major challenge was to bring local authorities together to agree on a common strategy and to develop a comprehensive plan of action. This could be done, for example, by the development of an extra curriculum on natural hazards or even by the inclusion of disaster education in the standard curriculum for primary schools.

Way Forward

There are several key issues concerning the development and implementation of tools for disaster education. An example that describes the tools within several types of learning processes is shown in Fig. 7. The tool for each type of learning process, such as learning process through lecture, experience, and presentation, differs from each other. The tools available for the learning process through lecture are generally in printed format, such as books (textbooks, comics/drawing books, and student working sheets/books), booklets, leaflets, and posters. Tools for the learning process through experience are the following: town/regional or community watching, interview, disaster drill, and various other action types of activities. Additionally, the workshop tool, such as action-oriented planning, is purposed for the learning process through presentation mode.

Another key issue is designing and tailoring disaster education tools for the specific (vulnerable) groups of people. There are growing needs in promoting disaster awareness through disaster education to the specific groups of people (children, women, community groups, etc.). For example, the design of awareness promotion tools that are appropriate, relevant, and adjusted to the specific learning needs (Clerveaux & Spence, 2009). Toward the development and implementation of tools for disaster education, it should be customized to the local context. Adaptation of a foreign disaster education tool into the local context should first be analyzed whether they are suitable at the targeted context. The issues that can be found and considered in the adaptation process are the following: (i) the physical–environmental differences, where different threats and environmental features exist, (ii) the social–cultural value differences, where it is possible

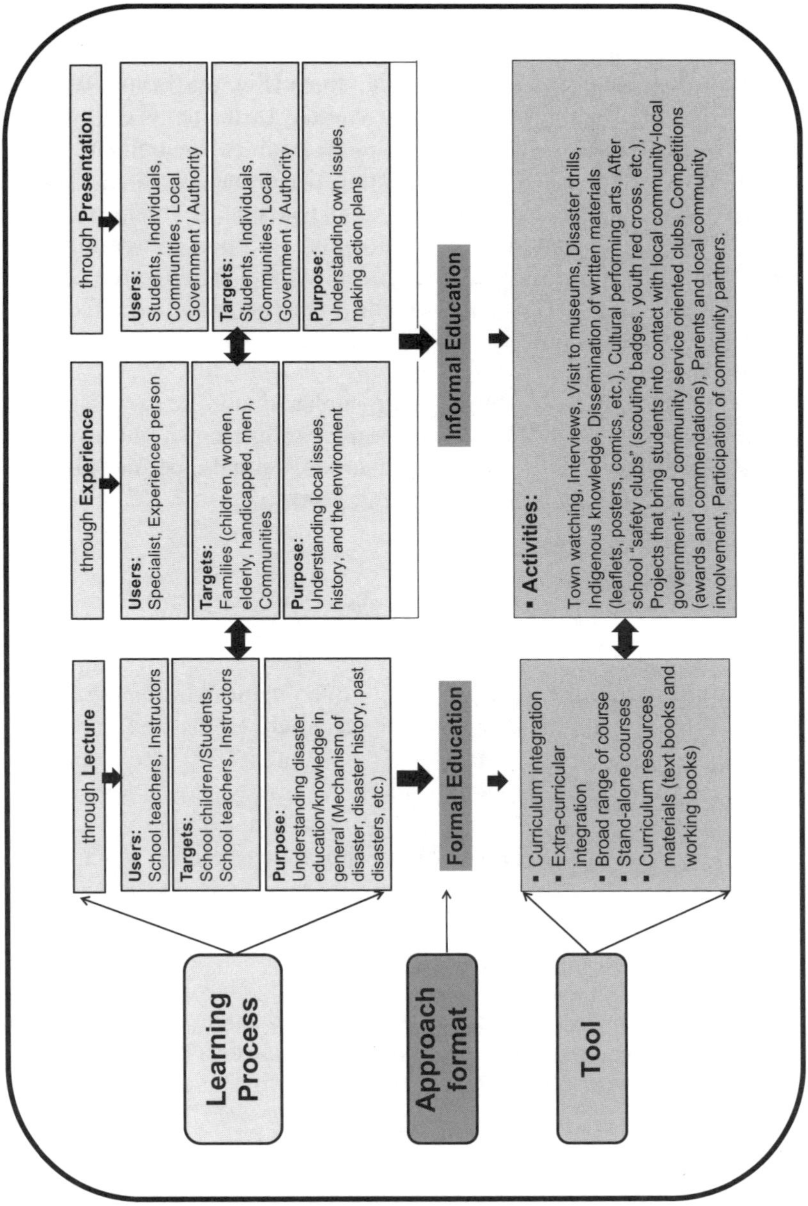

Fig. 7. Comprehensive Learning System and Its Tools for Disaster Education.

to integrate the disaster education tool into existing socioculture value, and (iii) the gap between the informal institution aspects applied from the original context to the new context, which should be reduced.

Another critical issue is in implementing, conveying, and communicating the tools to the end users. Education is not merely an academic exercise, but it is a knowledge transfer system to achieve sustainable development. In this regard, those to be entrusted with imparting the disaster information, for example to specific vulnerable groups of people such as children, should be adequately and appropriately trained not only in the content of such knowledge but also in relation to the methodologies of effective communication (Wisner, 2006). Therefore, in line with the essentials of disaster education, the tools on how to convey it determine the successfulness of disaster education.

Once the tools for disaster education are implemented, the next issue is on how those tools can be sustained over time. Long-term monitoring and evaluation is necessary. Last but not least, tools should be modified and flexible to the current demand and growing needs.

ACKNOWLEDGMENTS

The authors are thankful to the Japanese Government (Monbukagakusho: MEXT) scholarship support extended for this study to the first author.

REFERENCES

American Red Cross (2008). *Masters of disaster*. Available at http://www.redcross.org/disaster/masters/. Accessed on December 19, 2010.

Bhattia, S. (2006). *Safe schools in Uttar Pradesh: Earthquake resistant school buildings in Uttar Pradesh*. Report.

Clerveaux, V., & Spence, B. (2009). The communication of disaster information and knowledge to children using game technique: The disaster awareness game (DAG). *International Journal of Environment Resources*, *3*(2), 209–222.

Kikkawa, T. (1999). *Risk communication: Aiming at mutual understanding and better decision making (in Japanese)*, (p. 197). Tokyo Japan: Fukumura Press.

MANGEONAD (2008). Geo-risks reduction and awareness rising on natural hazards in primary schools in Nanggroe Aceh Darussalam, Indonesia. Newsletter for Community Based Disaster Risk Reduction (CBDRR) in Southeast Asia. Vol. 4, No. 3, p. 4. ADPC (PDRSEA), DIPECHO, UNESCAP.

Petal, M., & Izadkhah, Y. (2008). *Concept note: Formal and informal education for disaster risk reduction*. Available at http://www.riskred.org.activities/ddredislamabad.pdf. Accessed on December 19, 2010.

Shaw, R., & Takeuchi, Y. (2009a). *Town watching handbook for disaster education. Enhancing experiential learning*. European Union (EU), International Strategy for disaster risk reduction (ISDR), and Kyoto University-Japan.

Shaw, R., & Takeuchi, Y. (2009b). *Sustainable community disaster education in Saijo city and its effectiveness in landslide risk reduction*. Available at http://www.unisdr.org/prevention web/files/12062_07Shaw1.pdf. Accessed on December 19, 2010.

Talero, G. (2004). *Literature review environmental education and public awareness*. Available at http://worldfish.org/PPA/PDFs/Semi-Annual%20II%20English/2nd%20s.a%20eng_F2.pdf. Accessed December 19, 2010.

Team Safe-T (2008). *Team Safe-T for Schools. California partnership for safety and preparedness*. Available at http://www.teamsafe-t.org/schools.cfm. Accessed on December 20, 2010.

Wisner, B. (2006). *ISDR- Let our children teach us! A review of the role of Education and knowledge in disaster risk reduction*. Available at http://www.unisdr.org/eng/task%20force/working%20groups/knowledge-education/docs/Letour

Yoshida, Y., Takeuchi, Y., & Shaw, R. (2009). *Town watching as a useful tool in urban risk reduction in Saijo in urban risk reduction: An Asian perspective*, (pp. 189–205). Bingley, UK: Emerald.

CHAPTER 8

"*TSUNAGARU*": THE ESSENCE OF DISASTER EDUCATION

Rajib Shaw, Yukiko Takeuchi and Koichi Shiwaku

INTRODUCTION

While there are different approaches, tools, and target groups of disaster education, possibly the best starting point is children and students. Child-centered disaster risk reduction (DRR) has drawn significant attention in recent days (Plan, 2010). Child-centered DRR is a flexible rights-based approach combining child-focused (for children) and child-led (by children) activities with interventions geared toward bringing about change in community, local, and national duty bearers. It applies strategies such as awareness raising, capacity building, group formation, institutional development, research, and influencing and advocacy across a range of arenas. Analyzing the Hyogo Framework for Action (HFA) aims, Plan (2010) argues that child-centered DRR contributes to different HFA priorities as follows:

- Ensure that disaster risk reduction is a national and a local priority with a strong institutional basis for implementation – promoting and supporting children's rights.
- Involve children and young people to identify, assess, and monitor disaster risks and enhance early warning.
- Use knowledge, innovation, and education to build a culture of safety and resilience at all levels, because children are our future.

Disaster Education
Community, Environment and Disaster Risk Management, Volume 7, 153–162

ISSN: 2040-7262/doi:10.1108/S2040-7262(2011)0000007014

- Involve children and young people to reduce the underlying risk factors.
- Strengthen disaster preparedness for effective response at all levels, particularly at the community level, concentrating on children and young people.
- In their approach to disaster risk reduction, states, regional and international organizations, and other actors concerned should take into consideration the key activities listed under each of these five priorities and should implement them, as appropriate, to their own circumstances and capacities.

The same document (Plan, 2010) also argues that there are specific value additions of working with children as follows:

- Children are one of the most prominent vulnerable group
- Children can have better and holistic perspective of risks
- Children are effective in risk communication, especially to their families and communities
- Children can be innovative change agents
- Children can maximize the adaptive capacity needed to address climate change
- Children can change behaviors for more sustainable development
- Children are the leaders and decision-makers for tomorrow
- Children-centered DRR is a rights-based approach

Among the above arguments, one of the most important issues is the rights-based approach. Disasters are often seen as humanitarian affairs, and DRR is usually not linked to the "rights" issues in a proactive way. However, linking the child-centered DRR to a rights-based approach is new thinking, which needs further strengthening in its implementation through appropriate governance support. The "rights" referred to are the right for life, right to education, right to health, and right to participation. Built on varied legal systems and cultural traditions, the United Nations Convention of the Rights of the Child (UNCRC) is a universally-agreed set of non-negotiable standards and obligations. These basic standards – also called human rights – set minimum entitlements and freedoms that should be respected by governments. With these rights comes the obligation of both governments and individuals not to infringe on the parallel rights of others. These standards are both interdependent and indivisible; we cannot ensure some rights without – or at the expense of – other rights. Therefore, it is important and necessary to link DRR to children's rights.

DIVERSITY IN DISASTER EDUCATION

As a concept, child-centered DRR is important and essential, as mentioned above. However, this is not the sole purpose of disaster education, which varies from target groups and places. Disaster education has its variations based on hazards, governance, process and tools, and target groups.

Depending on the nature of the hazards, the disaster education approach should be different. Shaw, Shiwaku, Kobayashi, and Kobayashi (2004) argued the importance of earthquake disaster education compared to other types of disasters. Earthquake is a disaster that cannot be predicted. Thus, the only way to reduce the damages and losses caused by an earthquake is effective preparedness. However, in most countries, an earthquake occurs once in several years, which may vary from 10 to 50 or even 100 years. Thus, obviously the priority of earthquake is lower than other more frequent disasters such as floods, cyclones, and droughts. Consequently, in most countries, earthquake disaster remains within a dilemma of sudden occurrence (which needs preparedness as the only way to reduce losses) versus infrequent occurrences (which means the priority of preparedness becomes lower compared to other hazards). In contrast, hydro-meteorologic disasters such as floods and typhoons have different dimensions related to early warning and risk communication.

In the case of governance, different national and international frameworks on DRR can have differential impacts of the disaster educational activities. The Education for All (EFA) provides a framework of rights-based approach of education for all children. The Education for Sustainable Development (ESD) provides an umbrella concept to link different types of education, from development education to environment education and DRR education. The Hyogo Framework for Action (HFA) provides the framework for disaster education. E-HFA (Education in Hyogo Framework for Action) is a new concept which focuses on looking at the HFA through an education lens (Gwee, Shaw, & Takeuchi, 2011). The key point here is to focus on all five HFA priorities, in contrast to the traditional focus on HFA-3 for the education sector. However, in the implementation of E-HFA, there are variations in the national level (to be implemented by Ministry of Education) and in the local level (to be implemented by the Department of Education in city or local governments). A proper linkage of these activities on the national and local levels is required. In a recent study, Gwee, Takeuchi, Jet-chau, and Shaw (2011) identified 16 tasks under five HFA priorities, and have analyzed these tasks in a local government in Yunlin County in Taiwan. The study pointed out that out of 16 tasks,

6 tasks need further improvements, including: to prioritize DRR and allocate appropriate resources for DRR education; to strengthen early warning in the education sector through effective communication and dissemination mechanisms; to enhance dissemination of DRR information; and disaster recovery, to develop a recovery planning process that incorporates DRR, build on disaster preparedness capacities and mechanisms in the education sector, and assess disaster response preparedness capacities and mechanisms through strengthened planning. For other local governments, the priority will be different.

Educational processes and tools also have diversity. Sharma (2005) pointed out that the approach should be one of mainstreaming risk reduction in currently prevalent developmental or even relief and rehabilitation programs. Risk reduction calls for a change to safer development. As such, it forms part of the developmental domain, and has the chance of much wider and sustained impact through it. For this, it is important to understand local needs of working professionals/NGOs for translating them into a specific training curriculum. Small initiatives in remote places, whether taken by governments, developmental agencies, voluntary groups, or communities themselves, provide the most practical input for any lesson-learning exercise. Risk education thus needs to be based on case-teaching methodologies, the content for which will come from a repository of case studies, focusing on good practices collected from the field. Case studies may be documented in any medium suitable to field-level learning – these may include radio programs, street theatre, stories, photographs, songs, and chants. Effective risk reduction finds its roots in the simplification of complex issues and taking the science and technology of safe habitat planning to the people. Research on existing technology models, innovations and improvements, adaptation to local conditions and needs, and locally appropriate dissemination packaging are essential to make use of research and experience existing in the sector and transferring their benefits to communities who need them most. Sustainability and up-scaling are key issues for risk reduction interventions. International centers of excellence can be instrumental in up-scaling the program to a global level for delivery of trainings at any location and to any target audience through suitable networking with the closest knowledge center, NGO, and UN body. They may also offer credits for the certificate courses that may assist trainees in pursuing higher education at these centers.

The concept of "knowledge for all" is a starting point for this thought sequence. Knowledge is abundant in explicit forms such as books, but tacit knowledge such as that possessed by frontline field practitioners and local communities does not get explained, documented, or disseminated. The best method for tapping this tacit knowledge is through compilation of case studies of frontline situations and experiences. Knowledge banking and knowledge mapping are the means considered most appropriate for this purpose. A starting point for this will be to first map existing knowledge and conduct an audit of who is doing what and where. The Global Open Learning Forum for Risk Education (GOLFRE) is one of the innovative approaches to filtering up the field-based knowledge to practice (Sharma, 2005). The education initiative itself may start with a foundation course for field workers. There may be focal points in partner organizations for their respective geographic areas. A university-NGO partnership may be founded to implement such courses. In the longer term, the courses can be best run through an active enterprise between all partners: local NGOs, UN agencies, universities, specialized units, media, and government institutions.

KEY EMERGING MESSAGES FOR DISASTER EDUCATION

This book, through six key chapters, focuses on different aspects of disaster education and provides some emerging messages and directions for future disaster education. In chapter 2, Gwee et al. (2011) provided an innovative idea of looking at disaster education through an appropriate policy and governance structure, E-HFA (Education through Hyogo Framework for Action). The 16 tasks identified from the HFA Words into Action will be the future framework of governance system of the disaster education. Chapter 3 by Shiwaku and Fernandez (2011a) provides interesting case studies from different countries, and synthesizes the essentials of school disaster education as: incorporation of innovative approaches, balanced approach of school curriculum and extra curriculum activities, and linkage from the school to outside the school. The chapter also emphasizes that school education is not just by the teachers, but also different other stakeholders, including NGOs, universities, and local governments. Table 1 shows the key messages emerging from different chapters in the book.

Chapter 4 by Takeuchi, Mulyasari, and Shaw (2011) points out the diversity in the form of family and community. It shows that a rural community is different from an urban one, and accordingly, the family composition also shows its individual character, which defines specific family and community education. The chapter also suggests self-help,

Table 1. Summary of Key Messages of Disaster Education.

	Topics	Key Messages
Chapter 2	Disaster education policy	E-HFA as the key framework Link to other existing frameworks (EFA, ESD, LLL, CCE, and EE)
Chapter 3	School education	Innovative concepts and programs School curriculum and extra-curriculum Linking inside and outside schools
Chapter 4	Family and community education	Diversity of family and community types Knowledge-sharing system Participation in community groups
Chapter 5	Higher education	Field-based research and inter-disciplinary education Focus on processes and approaches Professional training/ job market creation
Chapter 6	Approaches	Linking school and community Promotion of disaster education Internet-based system
Chapter 7	Tools	Understanding scientific knowledge Quality control of education tools How to create interest in the KIDA model Link to indigenous knowledge

mutual-help and public-help differentiation and argues that a certain level of social association is important for family and community education. The next chapter by Shaw, Mallick, and Takeuchi (2011) shows the importance of higher education, and argues that the process is the most important for the higher education, rather than the curriculum. Inter-disciplinary education and field-based research system is required. Networking of universities helps to share and disseminate important lessons widely. Developing proper job markets for professionals, and appropriate training programs are an absolute need. Chapter 6 by Shiwaku and Fernandez (2011b) shows innovative approaches of disaster education citing different examples, and emphasizes that providing training to teachers, student's learning opportunities from the communities, and sharing information widely through the Internet are some of the successful approaches for effective disaster education. Chapter 7 by Mulyasari, Takeuchi, and Shaw (2011) suggests that there are different tools available for disaster education. Some of them are through lectures, some through experiences, and some through presentation. To enhance learning through all these, different tools are used, some through curriculum, and some through activities such as

Table 2. Alphabetical Connotation of Disaster Education.

A	Leads to ACTION
B	Should be BARRIER-FREE and gender-sensitive
C	Can cause a CHAIN REACTION from the school to the home and community
D	Helps in DISASTER RISK REDUCTION (DRR)
E	Is part of EDUCATION FOR SUSTAINABLE DEVELOPMENT (ESD)
F	Should be FLEXIBLE (can be formal, informal, and non-formal)
G	Can contribute in the dissemination of GOOD PRACTICES
H	Is one of the priorities in the HYOGO FRAMEWORK FOR ACTION (HFA)
I	Should be INNOVATIVE and INTERESTING
J	Is a JOINT EFFORT among many stakeholders
K	Propagates KNOWLEDGE of hazards, vulnerabilities, and coping capacities
L	Should be included in LIFELONG LEARNING (LLL)
M	Supports the attainment of the MILLENNIUM DEVELOPMENT GOALS (MDGs)
N	Benefits from existing and new NETWORKS of dedicated DRR supporters
O	Takes advantage of many OPPORTUNITIES in building a culture of safety
P	Is a continuous and iterative PROCESS
Q	Should be monitored and evaluated to maintain high QUALITY
R	Must be RESPONSIVE to the needs of the local audience
S	Is needed for enhanced SCHOOL SAFETY
T	May sometimes involve special TRAINING for capacity-building
U	Can contribute in UP-SCALING disaster risk awareness efforts in the community
V	Can be conducted by VOLUNTEERS
W	Will increase the understanding of WARNING SYSTEMS
X	Can overcome XENOPHOBIA and encourage affected communities to accept foreign assistance
Y	Should motivate YOUTH PARTICIPATION in DRR
Z	Can promote ZEAL among students and teachers alike toward a culture of preparedness

town watches, safety clubs, etc. Summarizing all these, Table 2 shows the important elements and connotation of disaster education using the alphabet.

"TSUNAGARU": THE LINK

"Tsunagaru" is a Japanese word that means linking. This word conveys the key essence of disaster education. As mentioned in the previous sections as well as earlier chapters of the book, the key of disaster education is linkage. The linkage is between school and community, between community and family, between different disciplines, between different stakeholders, between nature and human, between past and future, and so on. Thus, this single word represents the right direction and meaning of disaster education.

Fortuin and Bush (2010) present the concept of cross-boundary education and linking theory and practice with respect to sustainable development education. The same concept is equally important for the disaster education. The role of teacher is not just to "teach," but to facilitate the learning process. Teachers should stimulate the students by asking questions and providing tools rather than telling them exactly what to do. A workshop-based approach among the students promotes a culture of self-learning. Here, the teacher's role is "*tsunagu*," to make the link between their thought processes. The inter- and cross-disciplinary approach is also another link, where students from different basic disciplines interact among each other, and try to learn from each other. This is especially important and relevant for higher education. Sterling (2009) mentioned that "sustainability is not just another issue to be added to an overcrowded curriculum but a gateway to a different view of curriculum, of pedagogy, of organizational change, of policy and particularly of ethos." This is also an important essence of disaster education. It is difficult to create a new curriculum just for disaster education; rather it is more effective if the disaster education knowledge is interlinked into the existing curriculum. Here also, the concept of linking the curriculum and different disciplines are important. The same word "tsunagaru" is the key essence here for linking different disciplines.

In a stakeholder analysis of effective risk communication in southern India, Takeuchi and Shaw (2010) identified different stakeholders such as government departments, universities, NGOs, institutes, local communities, and private sectors. The key point was how to include disaster-related issues when addressing daily aspects and priorities of life. One way is to include development support such as the improvement of self-help capabilities. Another way is to focus on education. School has an important role in the social system. Teachers impart a lot of things (not only disaster education) to students in school. After school, students go home and talk to their families about what they learned. Therefore, linking different stakeholders is another target of risk communication. Here also, the same word "tsunagaru" expresses the appropriate meaning through stakeholder linkages.

Another aspect of disaster education is the link between humans and nature. Disasters are regarded as the interplay of nature and humans. The link between environment and disaster is prominent in the area where natural and social issues merge, and this is specifically prominent in the rural areas where most of the communities depend on agriculture and natural resources for their livelihood. Resource conservation, utilization, and management are the key issues, where education becomes extremely important. Environmental education is considered as the root of disaster education. This linkage becomes explicit through town watching or mountain watching or coastal watching (collectively called neighborhood watching), which is considered an important tool. "Tsunagaru" expresses the key lessons here through linking humans with nature through the educational process.

Disasters were present from the very inception of mankind. People and communities accumulated different experiences over time, and in many cases these experiences became important in coping with future disasters. The indigenous knowledge (IK) and its transferability play an important role in the education system (Shaw, Sharma, & Takeuchi, 2009). One of the major issues of IK is that it evolved as part of the survival process of the people and communities. Thus, IK is very much characterized by a multidisciplinary nature and is based on food, security, human and animal health, education, natural resource management, and various other community-based activities. The other issue of IK is its dynamic evolution. IK is the result of a continuous process of experimentation, innovation, and adaptation. It has the capacity to blend with knowledge based on science and technology and should therefore be considered complementary to scientific and technologic efforts to solve problems in social and economic development. The third point of IK is the challenge of documentation. In most cases, IK is orally transmitted and thus, the challenges of its implementation are not properly documented. The fourth important challenge is the classification of IK based on geographic, thematic, and organizational contexts and in many cases, there are overlaps. The IK is an important educational tool that links past to present and future, and in this context, the word "tsunagaru" conveys the time connotation of this link.

In summary, disaster education is a process-based approach that links school to communities and families, links humans to nature, links different disciplines, links theory to practice, and links indigenous knowledge and its transferability over time. Disaster education is a non-ending process and effort, and it needs the participation of different groups of stakeholders from students and teachers, to government, non-government organizations, the media, and private sectors.

ACKNOWLEDGMENTS

This chapter summarizes the key findings of earlier chapters, and therefore, discussions among all the authors at different stages were very useful and productive, and this is highly acknowledged.

REFERENCES

Fortuin, K. P. J., & Bush, S. R. (2010). Educating students to cross boundaries between disciplines and cultures and between theory and practice. *International Journal of Sustainability in Higher Education*, *11*(1), 19–35.

Gwee, Q., Shaw, R., & Takeuchi, Y. (2011). Disaster education policy: Current and future. In: R. Shaw, K. Shiwaku & Y. Takeuchi (Eds), *Disaster education*. UK: Emerald Publisher.

Gwee, Q., Takeuchi, Y., Jet-chau, W., & Shaw, R. (2011). Disaster education system in Yunlin county, Taiwan. *Asian Journal of Environment and Disaster Management*, *3*(2), 189–204.

Mulyasari, F., Takeuchi, Y., & Shaw, R. (2011). Implementation tools for disaster education. In: R. Shaw, K. Shiwaku & Y. Takeuchi (Eds), *Disaster education*. UK: Emerald Publisher.

Plan. (2010). *Child centered disaster risk reduction: Building resilience through participation* (48 pages). London, UK: Plan International.

Sharma, A. (2005). Pro-active risk education targeting field practitioners. In: R. Shaw & B. Rouhban (Eds), *Education for sustainable development: Towards effective disaster reduction and enhancing human security* (pp. 57–61). Kyoto, Japan: UNESCO-Kyoto University Publication.

Shaw, R., Mallick, F., & Takeuchi, Y. (2011). Higher education in disaster risk reduction; prospects and challenges. In: R. Shaw, K. Shiwaku & Y. Takeuchi (Eds), *Disaster education*. UK: Emerald Publisher.

Shaw, R., Sharma, A., & Takeuchi, Y. (2009). Indigenous knowledge and disaster risk reduction. In: R. Shaw, A. Sharma & Y. Takeuchi (Eds), *Indigenous knowledge and disaster risk reduction: From practice to policy* (pp. 1–14). Nova Publisher.

Shaw, R., Shiwaku, K., Kobayashi, H., & Kobayashi, M. (2004). Linking experience, knowledge, perception and earthquake preparedness. *Disaster Prevention and Management*, *13*(1), 39–49.

Shiwaku, K., & Fernandez, G. (2011a). Role of schools in disaster education. In: R. Shaw, K. Shiwaku & Y. Takeuchi (Eds), *Disaster education*. UK: Emerald Publisher.

Shiwaku, K., & Fernandez, G. (2011b). Innovative approaches in disaster education. In: R. Shaw, K. Shiwaku & Y. Takeuchi (Eds), *Disaster education*. UK: Emerald Publisher.

Sterling, S. (2009). Towards sustainable education. *in Environmental Scientists* (February), 19–21.

Takeuchi, Y., Mulyasari, F., & Shaw, R. (2011). Role of family and community in disaster education. In: R. Shaw, K. Shiwaku & Y. Takeuchi (Eds), *Disaster education*. UK: Emerald Publisher.